I0061091

Turbulent Flow and Boundary Layer Theory:

Selected Topics and Solved Problems

Authored by

Jafar Mehdi Hassan

Automotive Engineering Section, Mechanical Engineering Department,
University of Technology,
Iraq

Riyadh S. Al-Turaihi

Mechanical Engineering Department, College of Engineering/Department of Mechanical Engineering, Babylon University,
Babil – Iraq

Salman Hussien Omran

Ministry of Higher Education & Scientific Research/Private Higher Education Directorate,
Iraq

Laith Jaafer Habeeb

Mechanical Engineering, University of Technology,
Iraq

Alamaslamani Ammar Fadhil Shnawa

Polytechnica University of Bucharest,
Romania

Turbulent Flow and Boundary Layer Theory:
Selected Topics and Solved Problems

Authors: Jafar Mehdi Hassan, Riyadh S. Al-Turaihi, Salman Hussien Omran,
Laith Jaafer Habeeb, Alamaslamani Ammar Fadhil Shnawa

ISBN (Online): 978-1-68108-811-2

ISBN (Print): 978-1-68108-812-9

ISBN (Paperback): 978-1-68108-813-6

BENTHAM SCIENCE PUBLISHERS LTD.

End User License Agreement (for non-institutional, personal use)

This is an agreement between you and Bentham Science Publishers Ltd. Please read this License Agreement carefully before using the ebook/echapter/ejournal (**"Work"**). Your use of the Work constitutes your agreement to the terms and conditions set forth in this License Agreement. If you do not agree to these terms and conditions then you should not use the Work.

Bentham Science Publishers agrees to grant you a non-exclusive, non-transferable limited license to use the Work subject to and in accordance with the following terms and conditions. This License Agreement is for non-library, personal use only. For a library / institutional / multi user license in respect of the Work, please contact: permission@benthamscience.net.

Usage Rules:

1. All rights reserved: The Work is 1. the subject of copyright and Bentham Science Publishers either owns the Work (and the copyright in it) or is licensed to distribute the Work. You shall not copy, reproduce, modify, remove, delete, augment, add to, publish, transmit, sell, resell, create derivative works from, or in any way exploit the Work or make the Work available for others to do any of the same, in any form or by any means, in whole or in part, in each case without the prior written permission of Bentham Science Publishers, unless stated otherwise in this License Agreement.
2. You may download a copy of the Work on one occasion to one personal computer (including tablet, laptop, desktop, or other such devices). You may make one back-up copy of the Work to avoid losing it.
3. The unauthorised use or distribution of copyrighted or other proprietary content is illegal and could subject you to liability for substantial money damages. You will be liable for any damage resulting from your misuse of the Work or any violation of this License Agreement, including any infringement by you of copyrights or proprietary rights.

Disclaimer:

Bentham Science Publishers does not guarantee that the information in the Work is error-free, or warrant that it will meet your requirements or that access to the Work will be uninterrupted or error-free. The Work is provided "as is" without warranty of any kind, either express or implied or statutory, including, without limitation, implied warranties of merchantability and fitness for a particular purpose. The entire risk as to the results and performance of the Work is assumed by you. No responsibility is assumed by Bentham Science Publishers, its staff, editors and/or authors for any injury and/or damage to persons or property as a matter of products liability, negligence or otherwise, or from any use or operation of any methods, products instruction, advertisements or ideas contained in the Work.

Limitation of Liability:

In no event will Bentham Science Publishers, its staff, editors and/or authors, be liable for any damages, including, without limitation, special, incidental and/or consequential damages and/or damages for lost data and/or profits arising out of (whether directly or indirectly) the use or inability to use the Work. The entire liability of Bentham Science Publishers shall be limited to the amount actually paid by you for the Work.

General:

1. Any dispute or claim arising out of or in connection with this License Agreement or the Work (including non-contractual disputes or claims) will be governed by and construed in accordance with the laws of the U.A.E. as applied in the Emirate of Dubai. Each party agrees that the courts of the Emirate of Dubai shall have exclusive jurisdiction to settle any dispute or claim arising out of or in connection with this License Agreement or the Work (including non-contractual disputes or claims).
2. Your rights under this License Agreement will automatically terminate without notice and without the

Bentham Science Publishers Ltd.
Executive Suite Y - 2
PO Box 7917, Saif Zone
Sharjah, U.A.E.
Email: subscriptions@benthamscience.net

BENTHAM SCIENCE

CONTENTS

FOREWORD i

PREFACE ii

PART 1 TURBULENT FLOW

CHAPTER 1 FUNDAMENTALS OF TURBULENT FLOW 1

1. INTRODUCTION TO TURBULENT FLOW 1
1.1. Nature of Turbulent 2
1.1.1. Irregularity or Randomness 2
1.1.2. Diffusivity 2
1.1.3. Large Reynolds Numbers 3
1.1.4. Three – dim. Vorticity Fluctuations 3
1.1.5. Dissipation 4
1.1.6. Continuum 5
1.1.7. Turbulent Flow are Flows 5
1.2. Methods of Analysis 5
1.3. The Origin of Turbulent 6
1.4. Diffusion of Turbulent 7
1.5. Length Scales in Turbulent Flows 14

CHAPTER 2 TURBULENT TRANSPORT OF MOMENTUM 23
1. INTRODUCTION 23
2. THE REYNOLDS EQUATIONS 24
2.1. Correlated Variables 26
2.2. Equation for the Mean Flow 27
2.3. The Reynolds Stress 29
3. ESTIMATE OF THE REYNOLDS STRESS 30
4. REYNOLDS STRESS AND VORTEX STRETCHING 32
5. PRANDTL MIXING LENGTH THEORY 34
6. THE LOGARITHMIC – OVERLAP LAW 38
7. TURBULENT VELOCITY PROFILE 41
8. TURBULENT FLOW SOLUTION 42
8.1. Flow in Pipes 42
8.2. Flow between Parallel Plates 44
9. EFFECT OF ROUGH WALLS 45
10. DERIVATION OF REYNOLDS STRESS EQUATION OF MOTION FOR TURBULENT FLOW 47

CHAPTER 3 THE DYNAMICS OF TURBULENCE 52
1. INTRODUCTION 52
2. TURBULENT KINETIC ENERGY (TKE) 52
3. SOLUTION OF TKE 55

CHAPTER 4 TRANSIENT FLOW 63
1. DEFINITIONS 63
2. PRESSURE CHANGES CAUSED BY AN INSTANTANEOUS VELOCITY CHANGE 64
3. WAVE PROPAGATION AND REFLECTION IN A SINGLE PIPELINE 68
4. CLASSIFICATION OF HYDRAULIC TRANSIENTS 71
5. CAUSES OF TRANSIENTS 71
6. EQUATIONS OF UNSTEADY FLOW THROUGH CLOSED CONDUITS 72
6.1. Assumptions 72
6.2. Dynamic Equation 72
6.3. Continuity Equation 75
7. VELOCITY OF WATERHAMMER WAVES 78
8. METHODS FOR CONTROLLING TRANSIENTS 83
8.1. General 83
8.2. Available Device and Methods for Controlling Transients 84

8.3. Surge Tanks 84
8.4. Types of Surge Tanks 86
8.5. Air Chambers 87
8.6. Valves 88
SOLVED PROBLEMS 92

PART 2 TURBULENT BOUNDARY LAYER

CHAPTER 5 BOUNDARY LAYER 151
1. INTRODUCTION 151
2. DEFINITIONS 154
2.1. Displacement Thickness 154
2.2. Momentum Thickness θ 155
2.3. Kinetic Energy Factor 155
2.4. Shape Factor 155
3. THE SEPARATION OF A B.L. 155
4. PRESSURE DRAG 157
5. BOUNDARY LAYER THEORIES 158
5.1. Motivation 158
5.2. A Concise Result for Separately Quasinearly Subharmonic Functions 162
5.2.1. Laminar B.L 163
5.2.2. Turbulent B.L 163
5.3. Simulation Solution for Steady 2D. Flow 165
5.4. The Blasius Solution for Flat – Plate Flow 166
5.5. The Falkner – Skan Wedge Flows (General Solution) 168
5.6. Thwaites Method (Steady Flow) 172

CHAPTER 6 TURBULENT BOUNDARY LAYER 176
1. TURBULENT B.L. EQUATION 176
2. THE RELATIONS BETWEEN STRESSES AND PRESSURE GRADIENTS 179
2.1. Laminar Sub – Layer 181
2.2. The Inner Region of the Turbulent Layer 183
3. THE SKIN FRICTION COEFFICIENT $C_f(x)$ 185

CHAPTER 7 TRANSITION ZONE OF BOUNDARY LAYER 188
1. TRANSITION AND TURBLANCE 188
2. STABILITY ANALYSIS 189
3. TRANSITION ZONE 191
3.1. Conditions at Transition 191
3.2. Mixed B.L. Flow on a Flat Plate with Zero Pressure Gradient 191
4. METHODS OF BOUNDARY – LAYER CONTROL 195
4.1. Method of the Solid Wall 195
4.2. Acceleration of the Boundary Layer (Blowing) 196
4.3. Suction 196
4.4. Injection of Different Gas 197
4.5. Prevention of Transition by the Provision of Suitable Shapes 197
4.6. Cooling of the Wall 197
SOLVED PROBLEMS 199

REFERENCES 283

SUBJECT INDEX 284

FOREWORD

Most flows encountered in engineering practice are turbulent, and thus it is important to understand how turbulence affects wall shear. However, turbulent flow is a complex mechanism dominated by fluctuation, and despite tremendous amounts of work done in this area by researchers, turbulent flow still is not fully understood. Therefore, we must rely on experiment and the empirical or semi-empirical correlation developed for various situations. In laminar flow, fluid particles flow in an orderly manner along path lines, and momentum and energy are transferred across streamlines by molecular diffusion. In turbulent flow, the swirling eddies transport mass, momentum, and energy to other regions of flow much more rapidly than molecular diffusion, greatly enhancing mass, momentum and heat transfer. As a result, turbulent flow is associated with much higher values of friction, heat transfer, and mass transfer coefficients. The author with his collaboration has been teaching the subject of turbulent flow and the boundary layer for the past twenty years. In addition, this monograph is essential based on the lectures delivered. The lecture notes were prepared to be utilized by the postgraduate student as a part of their research work in the field of mechanical engineering (power generation). Also including about a hundred solved problems that have been given during the courses and examinations. From the experience gained throng useful class discussion and feedback, the notes were revised to improve the clarity and necessary explanatory nots were added during each teaching semester. The subject matter has thus been thoroughly tested in the classroom and found suitable. This book is a compilation and no claim is made of its originality. Acknowledgments are due and hereby made to all the authors whose work has been used in the preparation of this text. Finally, this book emphasizes the need for postgraduate engineers to acquire great efficiency in using the tool of the study and researches in the field of turbulent flow, boundary layers, *etc*.

M. I. Abu-Tabikh
Mechanical Engineering Department
University of Technology
Baghdad - Iraq
E-mails: 20004@uotechnology.edu.iq
mimatabikhg@gmail.com

PREFACE

The present book is focused on fundamental concepts of turbulent flow with boundary layer analysis. It is the outgrowth of our several years of teaching postgraduate courses in mechanical engineering department at University of Technology. A general introduction to turbulent flow is provided discussing flows turbulent that occurring in nature and engineering application. Also transient flow, methods for controlling transients, turbulent models and dynamic equations are explained for unsteady flow through closed conduits. In this book, all the basic concepts in turbulent flow are clearly identiifed and presented in a simple manner. with illustrative and practical examples. We have also attempted to make this book self-contained as much as possible; for example, materials needed from previous courses, such as equations, theory and engineering mechanics, are presented. Each chapter also has a set of questions and problems to test the student's power of comprehending the topics. Many of our colleagues and academic friends helped us by giving valuable suggestions on the structure and content of this text and these were instrumental in improving the quality and presentation of this book. We wish to express our profound gratitude and appreciation to all of them.

It is expected that the book will be a useful reference/text for professionals/students of engineering and including dynamic research and project consultants undertaking turbulent flow and boundary layer methods analysis. We have tried not only to give a comprehensive coverage of the various aspects of turbulent flow analysis but provided an exhaustive appendix on interest examples . These examples, along with the topics discussed, will, we believe, help both students and teachers in carrying out turbulent flow analysis and solving problems. In this book, all the basic concepts in turbulent flow are clearly identiifed and presented in a simple manner with illustrative and practical examples.

Any suggestions for improving the contents would be warmly appreciated.

CONSENT FOR PUBLICATION

Not applicable.

CONFLICT OF INTEREST

The author declares no conflict of interest, financial or otherwise.

ACKNOWLEDGEMENTS

Declared none.

Jafar Mehdi Hassan
Automotive Engineering Section
Mechanical Engineering Department
University of Technology
Iraq
E-mail: Jafarmehdi1951@yahoo.com

Riyadh S. Al-Turaihi
Mechanical Engineering Department
College of Engineering/Department of Mechanical Engineering, Babylon University, Babil
Iraq
E-mail: eng.riyadh.sabah@uobabylon.edu.iq

Salman Hussien Omran
Ministry of Higher Education & Scientific Research/Private Higher Education Directorate
Iraq
E-mail: asst.prof.salman@gmail.com

Laith Jaafer Habeeb
Mechanical Engineering
University of Technology
Iraq
E-mail: 20021@uotechnology.edu.iq

&

Alamaslamani Ammar Fadhil Shnawa
Polytechnica University of Bucharest
Romania
E-mail: ammar.fadhil88@yahoo.com

Part I

Turbulent Flow

CHAPTER 1

Fundamentals of Turbulent Flow

Abstract: Turbulent flow will be the subject of research during this period and near future and hundreds of papers and articles are being published every year.

In this chapter, we emphasize on the physics side of turbulent flow with the engineering application, nature of turbulent, methods on analysis, diffusivity of turbulence, length scales in turbulent flows, energy and vorticity and their relation with length scale.

Diffusive and convective length scales and their relation with boundary thickness at the laminar and turbulent region and a comparison between them are presented in details.

Keywords: Diffusivity, Energy and vorticity, Length scales, Nature of turbulent.

1. INTRODUCTION TO TURBULENT FLOW

Fluid particles moving very irregular path cause change in momentum from portion of fluid to another [1].

Most flows occurring in nature and engineering application are turbulent.

1. Boundary layer in the earth is turbulent.
2. Jet streams in the upper troposphere are turbulent.
3. Cumulate clouds are in turbulent motion.
4. Water current below the surface of oceans are turbulent.
5. Wakes of cars, ships, submarines and aircrafts are in turbulent motion.
6. Most combustion processes involve turbulent.
7. Flow of natural gas and oil in pipelines is turbulent.

Jafar Mehdi Hassan, Riyadh S. Al-Turaihi, Salman Hussien Omran, Laith Jaafer Habeeb, Alamaslamani Ammar Fadhil Shnawa

1.1. Nature of Turbulent

From smoke stack one can get some ideas about the nature of turbulent, however, it is very difficult to give a precise definition of turbulent flow. All one can do is list some of the characteristics of turbulent flow.

1.1.1. *Irregularity or Randomness*

Fig. (**1.1**) makes a deterministic approach to turbulent Problems impossible, instead, one relies on statistical methods.

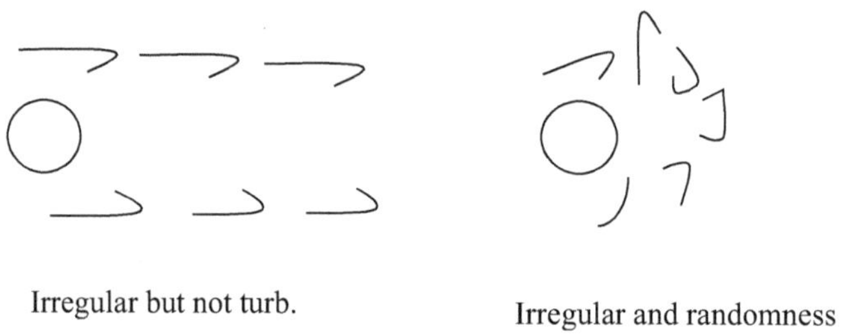

Fig. (1.1). Irregularity and randomness.

1.1.2. Diffusivity

It is the movement of mass due to molecular exchange (most important feature as far as applicant concerns.

The diffusivity of turbulence causes rapid mixing ⇒ increasing rate of momentum, heat mass transfer.

This done by ⇒ spreading velocity fluctuations through surrounding fluid (Turbulent)

If a flow pattern looks random but does not exhibit spreading of velocity, fluctuation it is not turbulent.

Diffusivity as far as application concerned help:

- Prevent B.L. separation.
- Increase H.T. rate.
- Resist flow in pipelines.
- Increase momentum transfer between winds and ocean current *etc*.

1.1.3. Large Reynolds Numbers

$$Re = \frac{Inertia\ force}{Viscus\ force} \gg 1 \quad \textbf{(1.1)}$$

- Turbulence often originates as instability of laminar flow if the Reynolds No. becomes too large.
- The instabilities are related to the interaction of viscous terms and nonlinear inertia terms in the equations of motion.
- Interaction is complex.
  Randomness + nonlinearity ⇒ Make eq. of turbulent nearly intractable.
- Theory suffers from powerful mathematical methods.
- This lack of tools ⇒ Trial and error.
- Turbulence research Both → challenging
  ↘ Frustrating

It is one of the principle unsolved problems in physics.

1.1.4. Three – dim. Vorticity Fluctuations

- Turbulence is ⇒ 3D + rotation.
- Applying force on particle means (motion + spin).
- Turbulence means ⇒ High level of fluctuation vorticity.

Therefore,

"Vorticity Dynamics applies important role in the dissipation of turbulent Flow"

- Vorticity, maintenance mechanism (Vortex stretching) is absent in 2D.

Turbulent flow always exhibits high levels of fluctuating vorticity.

1.1.5. Dissipation

Turbulent flows are always dissipative viscous shear stresses perform deformation work which increase the internal energy of the fluid at the expense of kinetic energy of the turbulence [2].

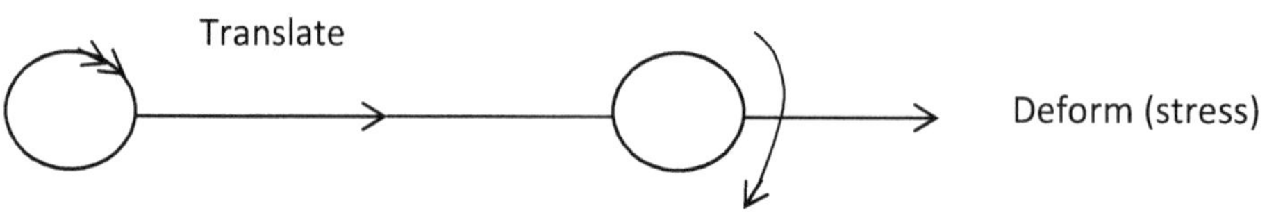

Fig. (1.2). Distinction between the random waves and turbulent flow.

The major distinction between random waves and turbulent is that waves are essentially nondissipative while turbulence is essentially dissipative as shown in Fig. (**1.2**).

Deformation means

$$V + V_d = \frac{\partial V}{\partial x}dx + \frac{\partial V}{\partial y}dy + \frac{\partial V}{\partial z}dz \quad \textbf{(1.2)}$$

$$\text{The gradient } \frac{\partial V}{\partial x} = \frac{\partial u}{\partial x} + \frac{\partial v}{\partial x}, \frac{\partial w}{\partial x} \quad \text{Turb. 3D} \quad \textbf{(1.3)}$$

$$\frac{\partial u}{\partial x} = \frac{1}{2}\left(\frac{\partial u}{\partial x} + \frac{\partial u}{\partial x}\right) + \frac{1}{2}\left(\frac{\partial u}{\partial x} - \frac{\partial u}{\partial x}\right) = \epsilon_{xx} + w_{xx} \quad \textbf{(1.4)}$$

$$\frac{\partial v}{\partial x} = \frac{1}{2}\left(\frac{\partial v}{\partial x} + \frac{\partial u}{\partial y}\right) + \frac{1}{2}\left(\frac{\partial v}{\partial x} - \frac{\partial u}{\partial y}\right) = \epsilon_{yx} + w_{yx} \quad \textbf{(1.5)}$$

$$\frac{\partial w}{\partial x} = \frac{1}{2}\left(\frac{\partial w}{\partial x} + \frac{\partial u}{\partial z}\right) + \frac{1}{2}\left(\frac{\partial w}{\partial x} - \frac{\partial u}{\partial z}\right) = \epsilon_{zx} + w_{zx} \quad \textbf{(1.6)}$$

If no energy is supplied, turbulence decays rapidly.

Turbulent needs continuous supply of energy to make up for these viscous losses.

1.1.6. Continuum

Turbulence is a continuum phenomenon governed by the equations of fluid mechanics [12].

Any smallest scales in turbulent is >> molecular length scale.

1.1.7. Turbulent Flow are Flows

Turbulence is not a feature of fluids but of fluid flows [17].

Turbulence is same fore ↗ Liquid
↘ Gas

- Eq. of motion is nonlinear.
- No general solution for N.S.E.
- No general difference.
- Turbulence depends on environment.

Therefore ⇒ No ingle theory for all kinds and types of flow is general.

1.2. Methods of Analysis

No general approach to solution of problems in turbulence exists.

- Statistical studies of the equations of motion always lead to a situation in which there are more unknowns than equations this called "Closure problem of turbulence theory ".
- The success of attempts to solve problems in turbulence depends strongly on the inspiration involved in making the crucial assumption.

On approach is use "Relation between stress rates of strain".
Another approach is "Phenomenological concept like eddy viscosity, mixing length".

The mathematical complexity of this work is so overwhelming that a discussion of it has to be left out of this book.

1.2.1. Dimensional Analysis

It is one of the most powerful tools, it relates to the relation between the " dependent and independent variables " it is useful if the turbulent depends on a few variables.

1.2.2. Asymptotic Invariance (Re Similarity)

Asymptotic invariance (Re similarity) and Turbulent flow are characterized by very high Reynolds No. *i.e.* Re $\rightarrow \infty$ or high Re means vanishing effect of molecular viscosity (with exception of very small scale of motion).

1.2.3. Local Invariance (Self – Preservation)

Means that " in simple flow geometry, the characteristics of the turbulent motion at same point in time and space appear to be controlled mainly by the immediate environment ".

If the turbulent time scales are small enough to permit adjustment to the gradually changing environment, it is often possible to assume that the turbulence is Dynamically similar everywhere i.e. wake.

Turbulent intensity in the wake is order of $\delta, \frac{\partial u}{\partial y}$

δ: the local width of the wake

$\frac{\partial u}{\partial y}$: average mean – velocity gradient across the wake.

1.3. The Origin of Turbulent

Channel flow *Re > 1500* Turbulence

B.L. flow with zero pressure gradient

$Re = \frac{\delta^* U}{\upsilon} = 600$ depends on surface

δ^*: displacement thickness.

- Turbulent flow is generally shear flow.

- Turbulence originate from an instability that causes vorticity, which subsequently becomes unstable, many wake flows becomes turbulent in this way.
- Turbulence cannot maintain itself but depends on its environment to obtain energy.
- Mathematical the details of transition from laminar to turbulent are poorly understood.
- Most theories based on linearization theory applicable for very small disturbance.
- For high distance ⇒ Turbulent → asymptotic theory.

1.4. Diffusion of Turbulent

- Turbulent motion ⇒ ability to transport or mix.
  Momentum, K.E, Heat, particles, moisture.
- Rate of transfer and mixing are several orders of magnitude greater than the rate due to molecular diffusion.

a) Diffusion in a problem with imposed length scale:

If there is no air, the room heat has to be distributed by molecular diffusion.

The distribution process is governed by the diffusion equation (1). Fig. (**1.3**) shows the control volume that was used.

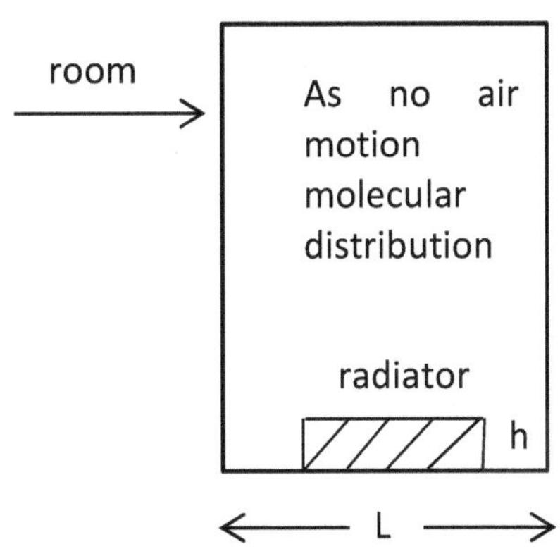

Fig. (1.3). The control volume.

$$\frac{\partial \theta}{\partial t} = \gamma \frac{\partial^2 \theta}{\partial x_i \, \partial x_i} \quad (1.7)$$

θ : is the temperature

γ : is the thermal diffusivity

assumed constant

Dimensionally

$$\frac{\Delta \theta}{T_m} \sim \gamma \frac{\Delta \theta}{L^2}$$

$$\therefore \ T_m \sim \frac{L^2}{\gamma} \Rightarrow \text{relates the time scale}$$

of moleculers $L + \gamma$

T_m : molecular time scale

$\Delta \theta$: Temp differance (Not related to the direction)

If

$$L = 5\,m \qquad \gamma = 0.2\ cm^2/sec$$

$$\therefore \ T_m \sim \frac{5^2 \times 10^4}{0.2 \times 3600} = 347\ hour.$$

We conclude that molecular diffusion is rather in effective in distributing heat through a room.

- In weak motion, such those generate by small density (buoyancy). For flow with length scale L and velocity scale u, the characteristic time scale T_t and is the r.m.s amplitude of velocity fluctuations in the room.

$$T_t \sim \frac{L}{u}$$

Where T_t can be determined only if u can be estimated.

→ suppose the radiator heats the air in its vicinity by $\Delta\theta\ k^{\circ} = 10\ k^{\circ}$

This causes a buoyant acceleration $= g\frac{\Delta\theta}{\theta}$

$$\sim \text{order of } 0.3\ m/s^2$$

This acceleration occurs only near the surface or radiator.

If its high $h = 0.1\ m$ hight (surface of radiator)

$$\therefore K.E = hg\,.\frac{\Delta\theta}{\theta} = 0.1 \times 0.3 = 0.03\ m^2/s^2$$

Since $\qquad K.E = \frac{1}{2}u^2 \qquad \left(\frac{1}{2}mu^2\right)\ unit\ mass\ m = 1$

$\therefore \qquad u = 0.24\ m/s$

$\qquad = 24\,cm/s \qquad$ for $L = 5\,m$

$$\therefore \qquad T_t = \frac{L}{u} = \frac{500}{24} = 20\ sec.$$

Therefore,

Diffusion by random motion apparently is very rapid compared to molecular diffusion.

Which is the inverse of the Peclet No.

$$\frac{T_t}{T_m} \sim \frac{L}{u}\cdot\frac{\gamma}{L^2} = \frac{\gamma}{uL} \qquad \textbf{(1.8)}$$

Where piclet No. $= \frac{\gamma}{uL}$

For gas $\gamma \approx \upsilon$ (Kinematic viscosity), the same order of magnitinde.

For air at $Re = 1500$ $\quad \frac{\upsilon}{\gamma} = 0.73$

$$\therefore \qquad \frac{T_t}{T_m} = \frac{\upsilon}{uL} = \frac{1}{Re} \tag{1.9}$$

$$\therefore \qquad \frac{1}{Re} \sim \frac{T_t}{T_m} \text{ rather than } \frac{Inertia\ force}{Viscous\ force}$$

Because, at high Re viscous effect tend to operate on small length scale than inertia effects.

Peclet No. $= \frac{\rho L u}{\Gamma}$ $\qquad \Gamma : \left.\begin{matrix}\text{diffusion}\\ \text{coefficient}\end{matrix}\right\}$ constant

The relative strengths of convective and diffusive transport.

$\gamma \simeq \upsilon \qquad \Gamma \simeq \mu$

Peclet identical to Reynolds No.

- Eddy Diffusivity

Equations of Turbulent is very complicate. we use effective diffusivity tend to treat turbulence as a property of a fluid rather than as a property of flow.

Diffusion of heat,

$$\text{By turbulent motion} \frac{\partial \theta}{\partial t} = K \frac{\partial^2 \theta}{\partial x_i \, \partial x_i}$$

Where K is diffusivity or eddy diffusivity.

To make this equation represent reality k must be chosen such that the time scale of the hypothetical turbulent diffiusion process is EQUAL to that of actual mixing process.

$$\therefore \text{ Time scale } T \sim \frac{L^2}{K} \text{ theoratical}$$

and actual time scale $T_t = \frac{L}{u}$

Equate

$$\frac{L^2}{K} \sim \frac{L}{u} \Rightarrow K = uL \qquad \textbf{(1.10)}$$

"This is dimensionally estimate which predict the numerical values of coefficient that may be needed".

The eddy diffusivity (or viscosity) K may be compare with kinematic viscosity.

$$\frac{K}{\gamma} \sim \frac{K}{\upsilon} \sim \frac{uL}{\upsilon} \gg 1 \quad \text{to be turbulent}$$

$$\approx Re \gg 1$$

∴ eddy diffusivity is much laser than the molecular diffusivity

We conclude that:

This particular Reynolds No. may also be interacting as a rate of

$$= \frac{\text{a pparent (turbulent)viscosity}}{\text{molecular viscosity}}$$

b) Diffusion in problem with an imposed time scale:

The B.L in the atmosphere is exposed to the rotation of the earth. The flows are accelerated by Coriolis force

$$\text{The angular velocity} \sim \frac{f}{2}$$

Laminar, its length and time scales are from diffusion equation.

$$\frac{\partial \theta}{\partial t} \simeq \gamma \frac{\partial^2 \theta}{\partial x_i \, \partial x_i} \qquad \textbf{(1.11)}$$

${L_m}^2 \sim \upsilon T$ Laminar B.L thickness at a latitude of 40 degrees, the value of f for a cartesion coordinate system whose z axis is parallel to the local vertical is about $10^{-4}\ sec^{-1}$.

$$\text{With } \upsilon = 0.15\ cm^2/s \text{ and } T = \frac{1}{f} = 10^4\ sec$$

This give $L_m = 40\ cm$ $\quad f =$ coriolios parameter

In reality as shown in Fig. (**1.4**), the atmospheric B.L is nearly always turbulent, a typical thickness is about 1 Km.

Hence substituted υ *by* K and $K = uL$

$$\therefore\ L_t^{\ 2} \sim K\,T \sim u\,L_t\,T$$

$$L_t \sim u\,T \sim \frac{u}{f}$$

u : eddies velocity

ℓ : largest eddy size

$L_t \approx B.L$ thickness

For example,

Characteristic velocity of Turbulent.

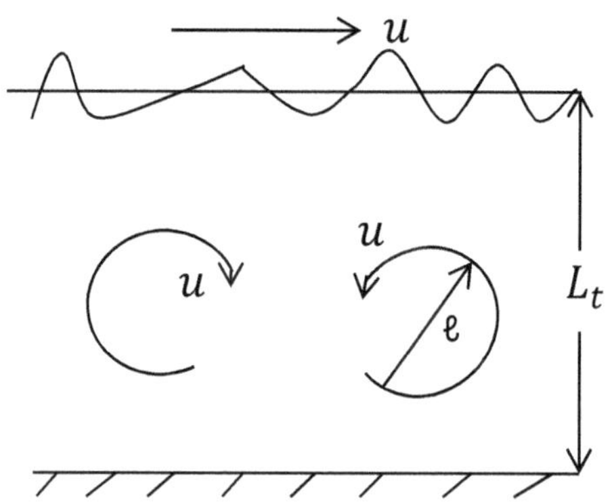

Fig. (1.4). Boundary layers.

$$u = \frac{1}{30} \text{ of mean wind speed.}$$

For wind speed = $10\ m/s$

$$u = 0.3\ m/s \quad with \quad T = \frac{1}{f} = 10^4\ sec$$

$$\therefore \quad L_t \sim u\,T \sim u\,.\frac{1}{f} = 3 \times 10^3\ m$$

or $L_t = 3 \times 10^3\ m$ at typical thickness as about $10^3\ m$.

To find the ratio between the thickness of the laminar and turbulent atmospheric B.L to be we have.

$$L_m{}^2 \sim \upsilon\,T \tag{1.12}$$

$$L_t \sim u\,T \tag{1.13}$$

$$T = \frac{1}{f}$$

$$\therefore \quad \frac{L_t}{L_m} \sim \frac{u\,T}{(\upsilon\,T)^{1/2}} \sim \frac{u}{f}\left(\frac{f}{\upsilon}\right)^{1/2} = \left(\frac{u^2 f}{f^2 \upsilon}\right)^{1/2} \tag{1.14}$$

$$= \left(\frac{u^2}{f\upsilon}\right)^{1/2} == \left(\frac{u^2}{\frac{u}{L_t}\upsilon}\right)^{1/2} = \left(\frac{uL_t}{\upsilon}\right)^{1/2}$$

$$\therefore \quad \frac{L_t}{L_m} = (Re)^{\frac{1}{2}} \tag{1.15}$$

Turbulent flow penetrates much deeper into the atmospheric than Laminar flow,

in this case $Re = 10^7$.

Note : Carioles = $2\,w\,\frac{\partial r}{\partial t}$ acceleration due to rotation

As a summary:

1. Flow with imposed length scales *Re* is proportional to the ratio of time scales.

$$\frac{1}{Re} \sim \frac{T_t}{T_m}$$

2. Flow with imposed time scale.

Re is proportional to the square of ratio of length scales

$$Re \sim \left(\frac{L_t}{L_m}\right)^2$$

Since the *Re* of most flows are large those relations clearly show that turbulent is a far more effective diffusion agent than molecular motion.

1.5. Length Scales in Turbulent Flows

For turbulent flow exists several length scales which bounded from above by the dimension of the flow field and bounded from below by the diffusive action of molecular viscosity [3].

a) Laminar B.L:

Of an incompressible fluid with constant viscosity, the N.S.E. are:

$$u_3 \frac{\partial u_i}{\partial x_j} = -\frac{1}{\rho}\frac{\partial P}{\partial x_i} + \upsilon \frac{\partial^2 u_i}{\partial x_j \, \partial x_j} \quad \textbf{(1.16)}$$

Inertia term viscous term

$$i = 1 \;\; x - comp \quad i = 2 \;\; y - comp \quad i = 3 \;\; z - comp$$

"The viscous terms can survive at high *Re* only by choosing a new length scale (ℓ) such that the viscous terms are the same order of magnitude on the inertia term". Suppose we have only one characteristic length L and velocity V.

$$\therefore \qquad \frac{U^2}{L} \text{ inertia} \qquad \upsilon\frac{L^2}{U} \text{ viscous}$$

If we take the ratio,

$$\frac{\text{inertia}}{\text{viscouse}} = \frac{V^2L^2}{L\,\upsilon\,U} = \frac{U\,L}{\upsilon} = Re \tag{1.17}$$

Increase inertia let *Re* increase,

For high *Re,* viscous effect must be negligible. However, B.C. or I.C. make it impossible.

For one length scale (L) means that viscous term is out of problem which is impossible (no B.L), therefore we have to define new length scale ℓ (associated with viscous effect) and its small length scale and in order of magnitude of inertia term.

$$\therefore \qquad \frac{U^2}{L} \sim \frac{U}{\ell^2} \tag{1.18}$$

$$\text{Therefore} \qquad \frac{U^2}{L} = \upsilon\,\frac{U}{\ell^2} \Rightarrow \frac{U}{L} = \frac{\upsilon}{\ell^2}$$

$$\text{or} \qquad \frac{\ell^2}{L^2} = \frac{U}{\upsilon L} \quad \text{dividing by L}$$

$$\frac{\ell}{L^2} = \left(\frac{U}{\upsilon L}\right)^{\frac{1}{2}} = \left(\frac{1}{Re}\right)^{\frac{1}{2}}$$

$$\therefore \qquad \frac{\ell}{L} = Re^{-\frac{1}{2}} \tag{1.19}$$

where ℓ is the transverse length scale. It represents the width (thickness) of B.L., as shown in Fig. (**1.5**).

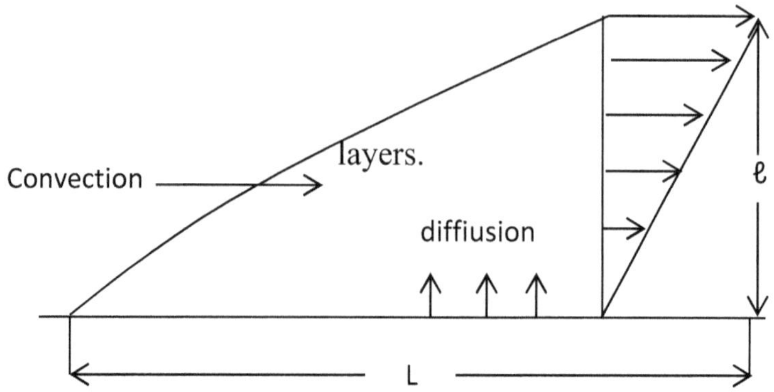

Fig. (1.5). Determine the thickness and the transverse length scale of the boundary.

Length scales, diffusion, and Convection in a Laminar B.L. over a flat plate.

We conclude that we need more than one characteristics length to deal with B.L. problem. May be need more than 2. Characteristics length to deal with Laminar flow *i.e.* diffusion length scale ℓ.

b) Turbulent B.L.

Fig. (**1.6**) shows the turbulent boundary layers schematic diagram.

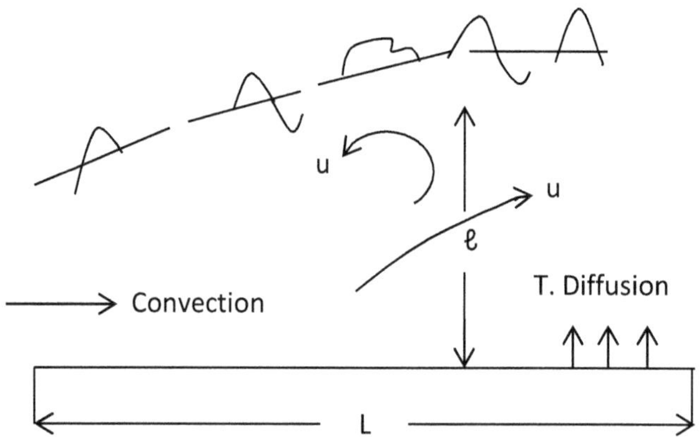

Fig. (1.6). Turbulent boundary layer.

$$\frac{d\ell}{dt} = u \tag{1.20}$$

We have,

Convective velocity u (fluctuation velocity)

Convective length L
Diffusion length ℓ
Stream velocity U
Convective time scale $\frac{\ell}{u}$
Diffusion time scale (turbulent) $\frac{\ell}{u}$
Equal time scale $\frac{L}{U} = \frac{\ell}{u}$

or $\frac{\ell}{L} = \frac{u}{U}$ this is order of 10^{-2} over a wide range of *Re,*

c) Laminar and Turbulent drag

If we compare $\frac{\ell}{L} = Re^{-1/2}$ Laminar
For experimental $\frac{u}{U} = 10^{-2}$ Turbulent

We find that "the velocity rapid growth of turbulent shear flows"

⇒ means large drag coefficient.

For steady laminar B.L in 2D flow on a plate, with length L, the drag per unit depth equal to the total rate of loss of momentum.

$$m \times \text{velocity} = \rho\, U\, \ell\, U \qquad D = \rho\, U^2 \ell$$

$$C_d = \frac{D}{\frac{1}{2}\rho\, U^2 L} \quad \Rightarrow \quad C_d = \frac{\rho\, U^2 \ell}{\frac{1}{2}\rho\, U^2 L} = 2\frac{\ell}{L} \tag{1.21}$$

$$\text{or} \quad C_d = 2Re^{-\frac{1}{2}} \qquad \text{a function of } Re.No.$$

But for Turbulent,

$$\text{momentum} = \rho\, U\, \ell\, .u \qquad D = \rho\, \ell\, U.u$$

$$\therefore \quad C_d = \frac{\rho\, \ell\, U.u}{\frac{1}{2}\rho\, U^2 L} = 2\frac{u\, \ell}{U\, L} \qquad \textbf{(1.22)}$$

$$\text{or} \quad C_d = 2\left(\frac{u}{U}\right)^2 \qquad \text{a function of velocity}$$

Experimental evidence show that:

Turbulent level $\frac{u}{U}$ varies very slowly with *Re* so that drag coefficient of turbulent B.L should by very much greater than the drag coefficient of laminar (See Fig. **1.7**).

$$\text{Laminer} \quad C_d = \frac{1.288}{Re^{0.5}}$$

$$\text{Turbulent} \quad C_d = \frac{0.72}{Re^{0.2}}$$

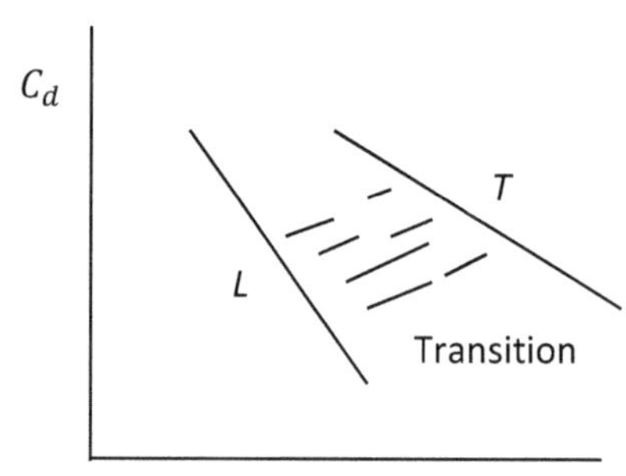

Fig. (1.7). The drag coefficient.

"Since small – scale motion tends to have small time scale"

- Small scale in turbulent, the large eddies do most of the transport of momentum to heat".
- Small scale fluctuation (motion) is produced by nonlinear, but viscous term prevent this scale by dissipating small – scale energy to heat.

On other hand, the small – scale of motion automatically adjusts itself to the value of viscosity.

∴ small scale motion depends on the rate at which it is supplied with energy by the scale motion & kinematic viscosity, (See Fig. **1.8**).

Or rate energy supply = rate of dissipation

"Kolmogorov's Universal equilibrium theory of small scale structure".

"The net rate of change should be small compared to the rate at which energy is dissipated".

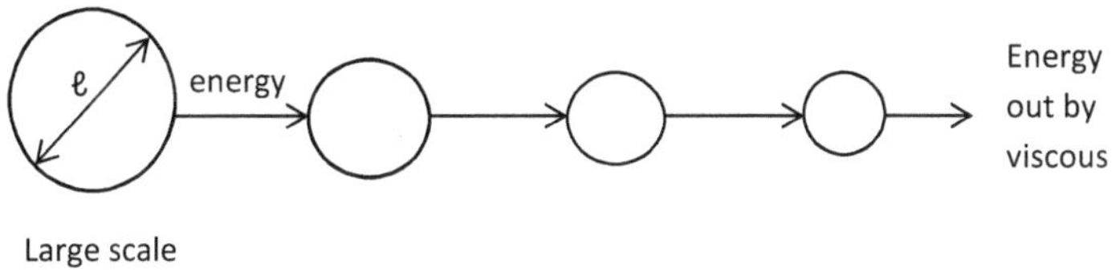

Fig. (1.8). Motion scales.

The discussion suggests that,

"The parameters governing the small – scale motion includes at least the dissipation rate/mass

$$\in \ (m^2/s^3) + \text{kinematic viscosity} \quad \upsilon \ (m^2/s)$$

With these parameters can form

Length, time, and velocity scales as follow;

$$\left.\begin{aligned} &Length && \eta = \left(\frac{\upsilon^3}{\in}\right)^{\frac{1}{4}} = 1 \\ &Time && \tau = \left(\frac{\upsilon}{\in}\right)^{\frac{1}{4}} = 1 \\ &Velocity && V = (\upsilon \in)^{\frac{1}{4}} = 1 \end{aligned}\right\} \text{these scale are refered to as the Kolmogrov' micro scale of length, time, velocity.}$$

The *Re*. No. formed with η, V is equal to one

$$Re = \frac{\eta\, U}{\upsilon} = 1$$

- Rate of supply energy to small – scale

$$u^2 \cdot \frac{u}{\ell} \sim \frac{u^3}{\ell} \qquad (m^2/s^3)$$

The energy is dissipated at a rate ∈ and should be equal to supply rate

$$\in \sim \frac{u^3}{\ell}$$

dissipated rate can be estimated from large scale dynamics, which do not involve viscosity.

Substituting the dissipation rate $\in = \frac{u^3}{\ell}$ in the length, time and velocity scale we get:

Scale relations (dimensionless parameters)

$$\left.\begin{aligned} \frac{\eta}{\ell} &= \left(\frac{u\ell}{v}\right)^{-\frac{3}{4}} \approx Re^{-\frac{1}{4}} \\ \frac{\tau\, u}{\ell} = \frac{\tau}{t} &= \left(\frac{u\ell}{v}\right)^{-1/2} = Re^{-\frac{1}{2}} \\ \frac{\mathrm{v}}{u} &= \left(\frac{u\ell}{v}\right)^{-\frac{1}{4}} \approx Re^{-\frac{1}{4}} \end{aligned}\right\} \qquad \textbf{(1.23)}$$

Means that, length, time and velocity scales of the smallest eddies are very much smaller than these of largest eddies [3].

→ Experimental film

As shown in Fig. (**1.9**).

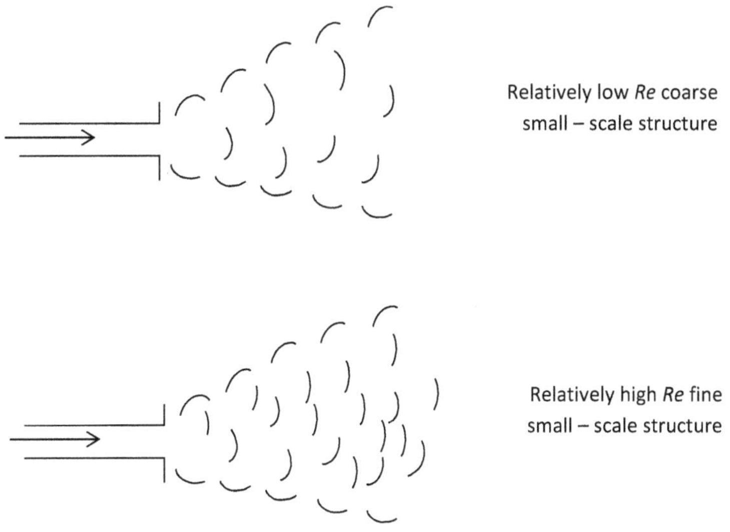

Fig. (1.9). Eddies types.

- Vorticity of small scale
- Dimension of vorticity $w \ rad/s \ \Rightarrow \ \frac{1}{sec}$ (frequncy)
- The vorticity of small – scale eddies $\sim \frac{1}{time\ scale\ \tau}$ we have time scale

$$\frac{\tau}{t} = Re^{-\frac{1}{2}} \quad \Rightarrow \ t \gg \tau$$

$$\text{Vorticity} \sim \frac{1}{\tau}$$

Vorticity of the small – scale eddies is very much larger than that of the large scale motion while small – scale energy is small compared to the large – scale energy, (Fig. **1.10**).

The typical of all turbulence

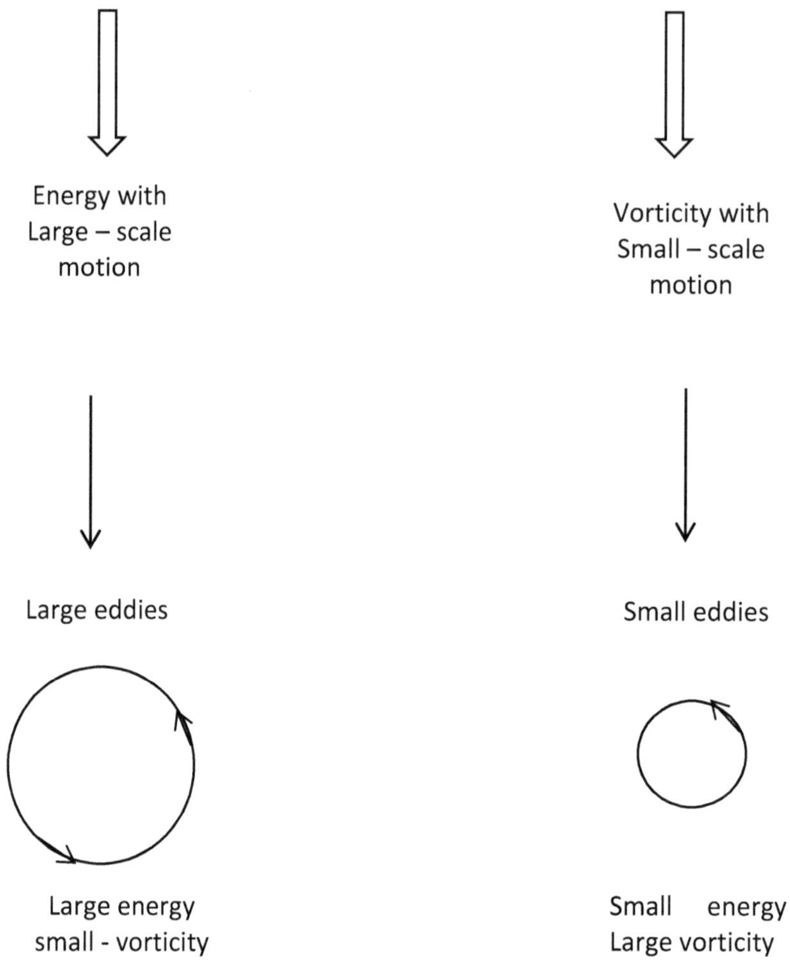

Fig. (1.10). Effect of motion on the velocity.

CHAPTER 2

Turbulent Transport of Momentum

Abstract: The Navier – stokes equations describe the transport of momentum in a viscous fluid. For a laminar flow, these N.S.E. can be solved directly, often to a high degree of accuracy.

For turbulent flow, things are more complex. The equations describe the instantaneous velocity components u, v, and w at every point in the flow. However, the nature of turbulence is such that there are very strong variation in these quantities over small distance. The time over which fluctuations in velocity occurs are likewise very small.

In this chapter a brief explanation and derivation of the N.S.E. for turbulent flow which they called "Reynolds stress".

Prandtl mixing length theory also presented to solve the "Reynolds stress" related to a length scale and velocity gradient. In addition, the velocity profiles for turbulent flow described throughout an experimental variation of inner – outer and overlap layer laws.

Keywords: Inner – outer and overlap layer, N.S.E, Reynolds stress, Velocity gradient.

2.1. INTRODUCTION

Turbulent consists of random velocity fluctuation, so that it must be treated with statistical methods. The statistical analysis does not need to be sophisticated at this stage: a simple decomposition of all quantities into mean values and fluctuations with zero mean [2, 3].

Decomposition:

$$u = U + u', \qquad \mathrm{v} = V + \mathrm{v}', \qquad w = W + w'$$

u, v, w are velocities measured at time and location
U, V, W are mean velocities (time average)

Jafar Mehdi Hassan, Riyadh S. Al-Turaihi, Salman Hussien Omran, Laith Jaafer Habeeb, Alamaslamani Ammar Fadhil Shnawa

u', v', w' are fluctuation velocities.

We interested in average quantizes,

Where,

$$U = \frac{1}{T}\int_{0}^{T} u\, dt$$

T is very large and u' fluctuation with zero mean *i.e.* $\overline{u'} = 0$, see Fig. (**2.1**).

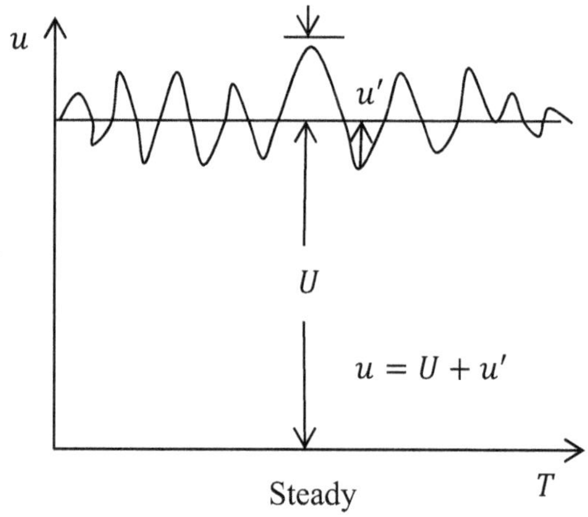

Fig. (2.1). Decomposition velocity.

Turbulent velocity fluctuation can generate large momentum fluxes between different parts of flow. A momentum fluxes can be thought of as a stress.

Turbulent momentum fluxes are commonly called Reynolds stresses. The momentum exchange mechanism superficially resembles molecular transport of momentum. The latter gives rise to the viscosity of a fluid, by analogy, the turbulent momentum exchange is often represented an eddy viscosity.

2.2. THE REYNOLDS EQUATIONS

The equations of motion of an incompressible fluid (N.S.E) are:

$$\frac{\partial \bar{u}_i}{\partial t} + \bar{u}_j \frac{\partial \bar{u}_i}{\partial x_j} = \frac{1}{\rho} \frac{\partial \bar{u}_i}{\partial x_j} \sigma_{ij} \tag{2.1}$$

$$\frac{\partial \bar{u}_i}{\partial x_i} = 0 \quad \text{(continuity eq.)} \tag{2.2}$$

Where $\bar{u}$ = instantaneous velocity value (x_i, t)

and $i = 1,2,3$ variation

$j = 1,2,3$ direction

i.e. " repeated indices mean summation "

if the fluid is Newtonian, the stress tensor

σ_{ij} is given by

$$\bar{\sigma}_{ij} = -\bar{P}\delta_{ij} + 2\mu\bar{s}_{ij} \tag{2.3}$$

Where

δ_{ij} is Kronecker delta, which

$$\delta_{ij} = \begin{bmatrix} 1 & \quad if\ i = j \\ 0 & \quad if\ i \neq j \end{bmatrix}$$

Example $i = 1$ $j = 1,2,3$ $\delta_{ij} = \delta_{xx} + \delta_{xy} + \delta_{xz}$
$i = 2$ $j = 1,2,3$ $\delta_{ij} = \delta_{yx} + \delta_{y} + \delta_{yz}$
$i = 3$ $j = 1,2,3$ $\delta_{ij} = \delta_{zx} + \delta_{zy} + \delta_{zz}$

P : hydrodynamic pressure

μ : dynamic viscosity (assumed constant)

s_{ij}: rate of strain defined by

$$\bar{s}_{ij} = \frac{1}{2}\left(\frac{\partial u_i}{\partial x_j} + \frac{\partial u_j}{\partial x_i}\right) \tag{2.4}$$

If eq. (2.4) substituted into eq. (2.1) we get,

$$\frac{\partial \bar{u}_i}{\partial t} + \bar{u}_j \frac{\partial \bar{u}_i}{\partial x_j} = -\frac{1}{\rho}\frac{\partial \bar{P}}{\partial x_i} + \upsilon \frac{\partial^2 \bar{u}_i}{\partial x_j \partial x_j}$$

Where υ is the kinematic viscosity $\left(\upsilon = \frac{\mu}{\rho}\right)$

2.2.1. Correlated Variables

Average of products are computed in the following way [1]:

$$\begin{aligned}\overline{u_i u_j} &= \overline{(U_i + u'_i)(U_j + u'_j)} \\ &= U_i U_j + \overline{u'_i u'_j} + \overline{U_i u'_j} + \overline{U_j u_i} \\ &= U_i U_j + \overline{u'_i u'_j}\end{aligned}$$

The terms consisting of a product of a mean value and a fluctuation vanish if they are averaged, because the mean value is a mere coefficient as far as the averaging is concerned, and the average of a fluctuating quantity is zero.

If $\overline{u'_i u'_j} \neq 0$, u'_i and u'_j are said to be correlated.

If $\overline{u'_i u'_j} = 0$ the two said uncorrelated.

Correlation coefficient for example channel flow $u'v'$ as shown in Fig. (**2.2**).

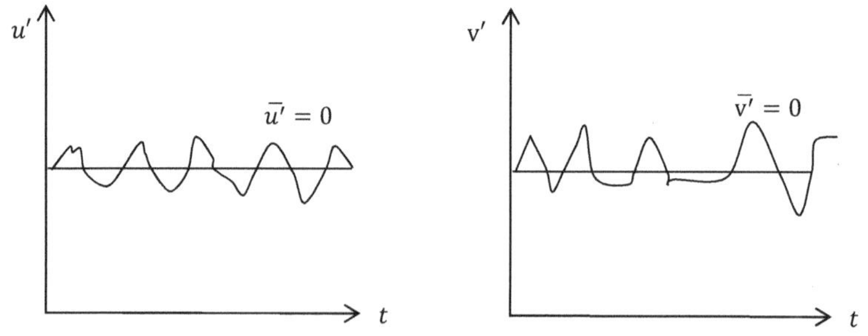

Fig. (2.2). Fluctuation velocity.

A measure for the degree of correlation between variables $u'v'$ is obtained by dividing $\overline{u'v'}$ by the square root of the product of the variances $\overline{u'^2}$ and $\overline{v'^2}$, this given correlation coefficient.

$$k_{uv} = \frac{\overline{u'v'}}{\sqrt{\overline{u'^2}}\sqrt{\overline{v'^2}}} \qquad \text{where} \qquad -1 \leq k_{uv} \leq 1$$

If $k = 0$ effect causing u' are unrelated to those causing v'.

Where $\overline{u'^2}$ is called a standard deviation or root – mean square (r.m.s) amplitude.

Where $\overline{u'^2} > 0$ and $\overline{v'^2} > 0$

2.2.2. Equation for the Mean Flow

The equation of motion is given by:

$$\frac{\partial u_i}{\partial t} + u_j \frac{\partial u_i}{\partial x_j} = \frac{1}{\rho} \frac{\partial}{\partial x_j} \sigma_{ij} \tag{2.5}$$

Substitute for u_i and taking the average and recall

$\frac{\partial u_i}{\partial t} = 0$ for incompressible fluid and the fluctuation

$\frac{\partial u'_i}{\partial x_j} = 0$ for turbulent fluctuation velocity (continuity equation).

We get,

$$U_j \frac{\partial U_i}{\partial x_j} + \overline{u'_j \frac{\partial u'_i}{\partial x_j}} = \frac{1}{\rho} \frac{\partial}{\partial x_j} \Sigma_{ij} \tag{2.6}$$

We may write,

$$\overline{u'_j \frac{\partial u'_i}{\partial x_j}} = \frac{\partial}{\partial x_j} \overline{\partial u'_i u'_j} \tag{2.7}$$

This is analogous to the convective term $U_j \frac{\partial U_i}{\partial x_j}$, it represent the "mean transport of fluctuating momentum by turbulent velocity fluctuating".

If u'_i and u'_j are uncorrelated, there would be no turbulent momentum transfer.

Experience shows that momentum transfer is a key feature of turbulent motion.

This means that eq. (2.7) is "the exchanges momentum between the turbulent and the mean flow".

Turbulent fluctuation are always 3D even the mean flow structure is 2D that is:

$$w' \neq 0 \qquad \frac{\partial}{\partial z}() = 0 \qquad w = 0$$

Consequently $w' \neq 0$ for any turbulent flow however, for

$$k_{uw} = \frac{\overline{u'w'}}{\sqrt{\overline{u'^2}}\sqrt{\overline{w'^2}}}$$

$$k_{vw} = \frac{\overline{v'w'}}{\sqrt{\overline{v'^2}}\sqrt{\overline{w'^2}}}$$

The physical effects causing w' are physically unrelated to those causing u' and v' on the average hence,

$k_{uw} = k_{vw} = 0$ for two – dim. Flow

Or

The turbulent motion can be assumed as an agency that produces stress in the main flow.

If all stresses put together, the equation of motion becomes,

$$U_j \frac{\partial u_i}{\partial x_j} = \frac{1}{\rho} \frac{\partial}{\partial x_j} \left(\Sigma_{ij} - \rho \, \overline{u'_i u'_j} \right) \tag{2.12}$$

viscous stress — turbulent stress

If we substitute for Σ_{ij} and s_{ij} obtained we get,

$$U_j \frac{\partial U_i}{\partial x_j} = -\frac{1}{\rho} \frac{\partial P}{\partial x_j} + \upsilon \frac{\partial^2 U_i}{\partial x_j \partial x_j} - \frac{\partial}{\partial x_j} \overline{u'_i u'_j} \tag{2.13}$$

Reynolds momentum equation for mean flow.

2.2.3. The Reynolds Stress

The contribution of the turbulent motion to the mean stress tensor is designated by [2, 3]:

$$\tau_{ij} \equiv -\rho \, \overline{u'_i u'_j}$$

The Reynolds stress is symmetric:

$$\tau_{ij} = \tau_{ji}$$

Or,

$$\tau_{ij} = -\rho \begin{bmatrix} \overline{u'^2} & \overline{u'v'} & \overline{u'w'} \\ \overline{u'v'} & \overline{v'^2} & \overline{v'w'} \\ \overline{u'w'} & \overline{v'w'} & \overline{w'^2} \end{bmatrix}$$

It can be seen the diagonal components of τ_{ij} are normal stresses (pressures), their values are, $\rho u'^2, \rho v'^2, \rho w'^2$. In many flows, these normal stresses contribute little to the transport of mean momentum. The off – diagonal components of τ_{ij} are shear stresses, the play a dominant role in the theory of mean momentum transfer by turbulent motion.

If $\overline{u'_i u'_j} \neq 0$ there is a force on fluid or there is a stress on fluid.

If we wish to solve for $U_i(u, \mathrm{v}, w$ and $P)$ in addition for these unknown we have the additional six independent unknowns,

$$\overline{u'^2}, \overline{\mathrm{v}'^2}, \overline{w'^2}, \overline{u'\mathrm{v}'}, \overline{\mathrm{v}'\mathrm{w}'}, \qquad \overline{\mathrm{u}'\mathrm{w}'}$$

Even for the very simple case of fully developed channel flow where,

$$\mathrm{v}' = \mathrm{w}' = \overline{\mathrm{u}'\mathrm{w}'} = 0 \quad \text{and} \quad \frac{\partial \overline{u'^2}}{\partial x} = 0$$

The equation for U retains one unknown.

Reynolds stress in addition to the mean velocity specially

$$\frac{1}{\rho}\frac{\partial P}{\partial x} = \upsilon \frac{\partial^2 U}{\partial y^2} - \frac{\partial \overline{u'\mathrm{v}'}}{\partial y}$$

The pressure of $\overline{u'\mathrm{v}'}$ prevents an analytical solution for $U_{(y)}$.

2.3. ESTIMATE OF THE REYNOLDS STRESS

We have seen that molecular transport can be interpreted fairly easily in terms of the parameters of molecular motion. It is very tempting to apply a similar heuristic treatment to turbulent transport. We again use a pure shear flow as a basis for our discussion. The rates of turbulent momentum transfer to be:

$$\tau_{12} \equiv -\rho\, \overline{u'_1 u'_2} \qquad \text{or} \qquad \tau_{xy} \equiv -\rho\, \overline{u'\mathrm{v}'}$$

The existence of Reynolds stress requires that the velocity fluctuations $u'\mathrm{v}'$ be correlated in shear flow (Fig. **2.3**)

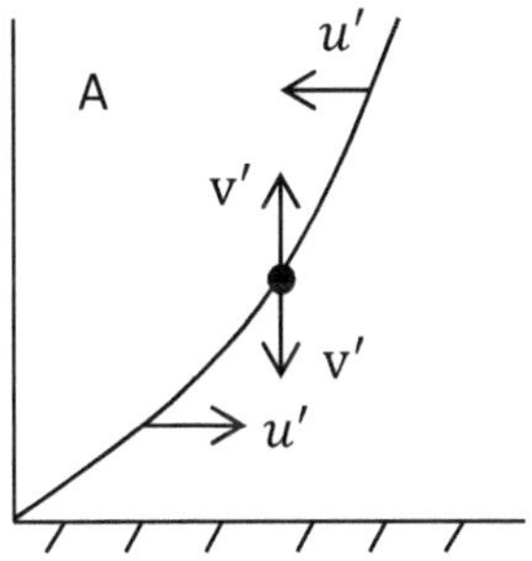

Fig. (2.3). Velocity fluctuations in shear flow.

At (A) $v' > 0 \quad u' < 0 \qquad \therefore u'v' < 0$

$v' < 0 \quad u' > 0 \qquad \therefore u'v' < 0$

Shear flow on flat plate (Fig. **2.4**)

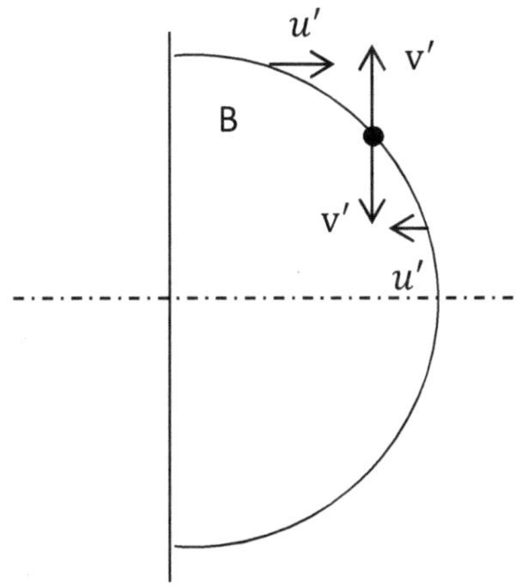

Fig. (2.4). Velocity fluctuations in shear flow on flat plate.

At (B) $v' > 0 \quad u' > 0 \qquad \therefore u'v' > 0$

$v' < 0 \quad u' < 0 \qquad \therefore u'v' > 0$

Shear flow in pipes.

At center line

$$v' > 0 \quad u' = 0 \qquad u'v' = 0$$
$$v' < 0 \quad u' = 0 \qquad u'v' = 0$$

u' increases close the wall (effect of viscosity).

The above estimation is the best way to find the correlation between u', v'.

2.4. REYNOLDS STRESS AND VORTEX STRETCHING

The energy of the eddies has to be maintained by shear flow, because they are continuously losing energy to smaller eddies.

Eddies on the other hand, need shear to maintain their energy, the most powerful eddies thus are those that absorb energy from the shear flow more effectively than others.

Experimental evidence suggests that the eddies that are more effective than most in maintaining the desired correlation between u' and v' and in extracting energy from the mean flow are " Vortex " whose principle axis is roughly aligned with that of the mean strain rate. Such eddies are illustrated as follows in Fig. (**2.5**).

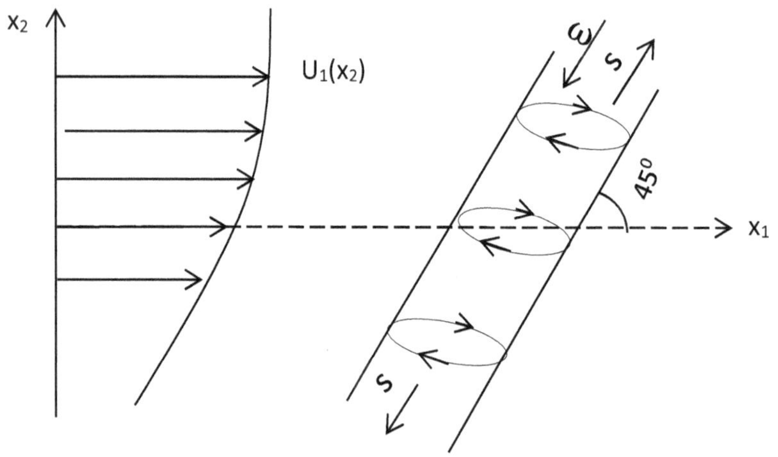

Fig. (2.5). Eddies formation.

The energy transfer mechanism for eddies of this kind is believed to be associated with vortex stretching *i.e.*

$$\frac{Dw}{Dt} = (w.\nabla)\mathrm{v} + \upsilon\nabla^2 w$$

and $w = \xi_i + \eta_j + \zeta_k$

vortex: the flow becomes vertical in the presence of velocity gradient (*i.e.* in B.L.) … it means that the B.L. is fundamentally a region of concentrated vorticity as shown in Fig. (**2.6**).

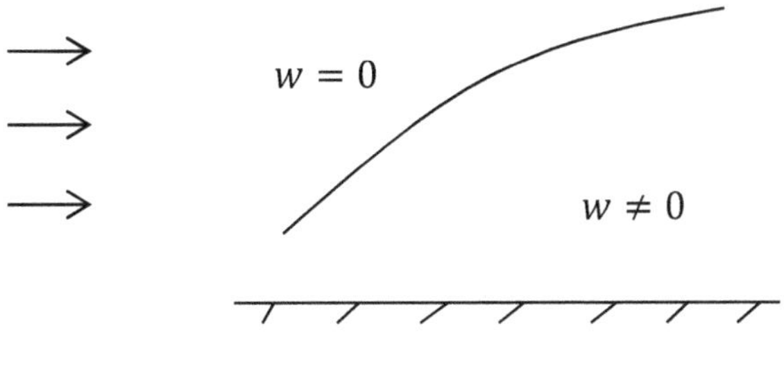

Fig. (2.6). Velocity gradient effect on boundary layers.

This makes it much easier to understand the behavior of a B.L. starting from rest, see Fig. (**2.7**).

The term $(w.\nabla)\mathrm{v}$ is termed vortex stretching or production, because,

1. Amplify or attenuate the vorticity by the elongation (or shortening) of vortex tube element (conservation of angular momentum).
2. Produce x- component vorticity w_x by reorientation the vortex tube which originally contained by any y- component vorticity.

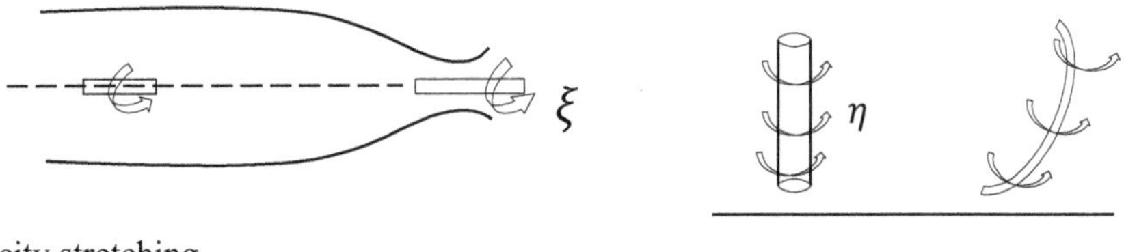

Fig. (2.7). Vorticity stretching.

Note: Discussion applies only to shear flow, otherwise its different.

2.5. PRANDTL MIXING LENGTH THEORY

The apparent shear stress in turbulent flow is expressed by:

$$\tau = (\mu_\ell + \mu_t)\frac{du}{dy} \tag{2.10}$$

Where μ_ℓ = viscosity of fluid due to laminar flow

μ_t = viscosity of fluid due to turbulent flow

Prandtl introduced a mixing coefficient, A_t for the Reynolds stress in turbulent flow, by putting

$$\begin{aligned} \tau_t &= -\rho\,\overline{u'\mathrm{v}'} \\ &= A_t\frac{du}{dy} = \rho \in \frac{du}{dy} \end{aligned} \tag{2.11}$$

$\in$= eddy viscosity (kinematic viscosity for turbulent flow)

Under this assumption that, A_t is not property of fluid like μ, but depends itself on the mean flow velocity U.

To explain Prandtl theory we consider the following [1]:

- The pressure gradients on fluid which accompany the adding motion have no effect on the mean transfer of the momentum.
- The particle of fluid is displaced a distance ℓ before its momentum is change by new environment.

A simple case of parallel incompressible flow in which the velocity varies only in y – dir. The principle of flow is assumed parallel to the x – axises and we have

$$u = u_{(y)},\ \ \mathrm{v} = 0,\ \ w = 0 \qquad \text{(channel flow)}$$

For this case only the shear stress,

$$\tau_{12} = A_t \frac{du}{dy} \quad \textbf{(2.12)}$$

The mechanism of the motion in Prandtl theory is as follows, (see Fig. **2.8**).

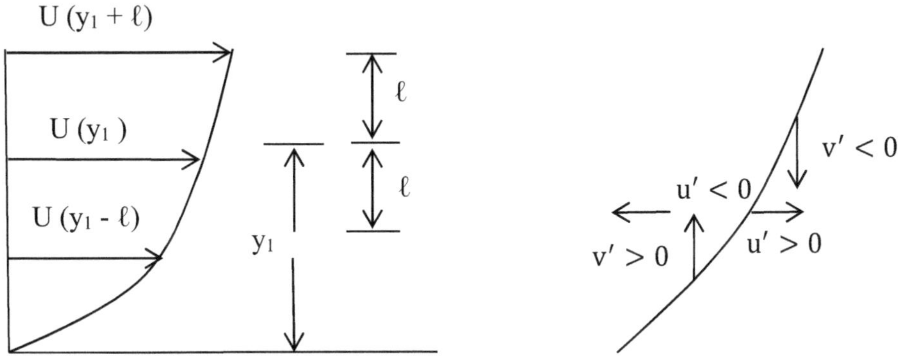

Fig. (2.8). Motion mechanism.

$u = U + u'$ if we neglect the effect of u' and assume that the mean velocity U remain constant during the transverse motion of a lump of fluid.

Based on this assumption "A lump of fluid, which comes from a layer at $(y_1 - \ell)$ and has velocity $U(y_1 - \ell)$ is displaced over a distance ℓ in the transverse direction as shown".

This distance is known as Prandtl mixing length.

Its velocity in new lamina at y_1 is smaller than the velocity prevailing there

∴ The difference in velocity is then:

$$\Delta u_1 = U(y_1) - U(y_1 - \ell) \approx \ell \left(\frac{dU}{dy}\right)_1 \quad \textbf{(2.13)}$$

Similarly, a lump of fluid which arrives at y_1 from lamina at $(y_1 + \ell)$ possesses a velocity which exceeds that around it, the difference being:

$$\Delta u_2 = U(y_1 + \ell) - U(y_1) \approx \ell \left(\frac{dU}{dy}\right)_2 \quad \textbf{(2.14)}$$

The velocity difference caused by the transverse motion can be regarded as the "turbulent fluctuation velocity at y_1"

Time – average of the absolute value of the fluctuation.

$$|\bar{u'}| = \frac{1}{2}(|\Delta u_1| + \Delta|u_2|)$$
$$= \ell\left|\frac{dU}{dy}\right| \qquad \textbf{(2.15)}$$

This equation leads to the following physical interpretation of the mixing length. "The mixing length is that distance in the transverse direction which must be converted by a lump of fluid particles travelling with its original mean velocity in order to make the difference between its velocity and the velocity in new lamina EQUAL to the mean transverse fluctuation in turbulent flow" [15].

The same argument can be implies that the transverse component v' is the same order of magnitude as u' and we put

$$|\bar{\mathrm{v}'}| = const.\,|\bar{u'}| = const.\,\ell\left(\frac{dU}{dy}\right)_1$$

To find expression for the shearing stress, the average u', v' is different from zero, and negative.

Hence,

$$\overline{\mathrm{v}'u'} = -\,c\,|\bar{u'}||\mathrm{v}'| \quad \text{with} \quad 0 \ll c < 1$$

$$\therefore\ \mathrm{v}'u' = -\,c\ell^2\left(\frac{dU}{dy}\right)^2$$

Or,

$$\overline{\mathrm{v}'u'} = -\,\ell^2\left(\frac{dU}{dy}\right)^2 \qquad c = 1 \qquad \textbf{(2.16)}$$

Shear stress can be written as,

$$\tau_t = -\rho u'v' = -\rho \ell^2 \left(\frac{dU}{dy}\right)^2$$

$$\text{or} \quad \tau_t = \rho \ell^2 \left(\frac{dU}{dy}\right)^2 \qquad \textbf{(2.17)}$$

taking into account that the sign of τ_t must change with that of $\left(\frac{dU}{dy}\right)$, it is found that it is more correct to write,

$$\tau_t = \rho \ell^2 \left|\frac{dU}{dy}\right| \left(\frac{dU}{dy}\right) \qquad \textbf{(2.18)}$$

This is Prandtl mixing – length hypothesis comparing this equation with equation (2).

$$\tau_t = \rho \in \left(\frac{dU}{dy}\right)$$

The expression for the eddy viscosity $\in$:

$$\in = \ell^2 \left|\frac{dU}{dy}\right|$$

Or,

$$\mu_t = \rho \ell^2 \left|\frac{dU}{dy}\right|$$

we are still faced with the proposition of assigning a value to ℓ, but simple dimensional reasoning can be useful.

Prandtl reasoned that:

a) Region not too distance from the wall (inner region)
   $\ell \sim y$ or $\ell = ky$

   where k is constant $\simeq (0.25 \rightarrow 0.41)$ von – Karman constant.

   This is not valid in the viscous sublayer (we discuss this region later).

b) Outer region: The experimental data shows that,

$$\ell \sim \delta \qquad (\delta \quad \text{B.L. thickness})$$

and ℓ becomes unimportant for y/δ_{99} greater than about 0.7.

∴ The outer region can be quite approximated by,

$$\ell = \lambda\, \delta_{99}$$

Where,

λ is typically about 0.085.

2.6. THE LOGARITHMIC – OVERLAP LAW

The turbulent shear stresses distribution of τ_{Lam} and τ_{turb} shows in Fig. **(2.9)** from typical measurement across a turbulent shear layer near the wall. Laminar shear is dominating near the wall (the wall layer), and turbulent shear dominate in the outer layer. There is an intermediate region called " Overlap Layer " [5].

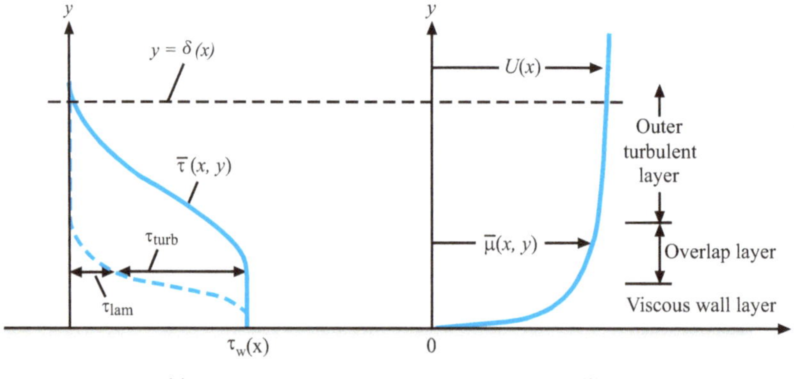

Fig. (2.9). Typical velocity and shear distribution in turbulent flow near a wall (**a**) shear (**b**) velocity [5].

Therefore,

1. Wall layer: Viscous shear dominates.
2. Outer layer: turbulent shear dominates.
3. Overlap layer: both types of shear important.

In the outer layer τ_{turb} is two or three orders of magnitude greater than τ_{Lam} and *viceve rsa* in the wall layer.

Let τ_w be wall shear stress and δ, U represents the thickness and velocity at the edge of the outer layer, $y = \delta$.

For the wall layer, Prandtl deduced that u must be independent of the shear – layer thickness,

i.e. $u = f(\mu, \tau_w, \rho, y)$

By dimensional analysis this is equivalent to,

$$u^+ = \frac{u}{u^*} = F\left(\frac{yu^*}{v}\right) \quad \text{where} \quad u^* = \sqrt{\frac{\tau_w}{\rho}} \tag{2.19}$$

Eq. (2.19) is called the law of the wall and u^* is termed the "Friction velocity" because it has dimensions (LT^{-1}), although it is not actually velocity.

Subsequently, Ka'rma'n deduced that u in the outer layer is independent of molecular viscosity but its deviation from the stream velocity U must be depend on layer thickness δ and the other properties.

$$(U - u)_{outer} = g(\delta, \tau_w, \rho, y)$$

Again, by dimensional analysis we rewrite this as,

$$\frac{\overline{U-u}}{u^*} = G\left(\frac{y}{\delta}\right) \tag{2.20}$$

Eq. (2.20) is called the "velocity defect law" for outer layer.

Both the wall law equation (2.19) and the defect law (2.20) are found to be accurate for wide variety of experimental turbulent duct and boundary layer flows. They are different in the form yet they must overlap smoothly in the intermediate layer. 1937 C.B. Millikan show that this can be true only if the overlap – layer varies logarithmically with:

$$\frac{u}{u^*} = \frac{1}{k}\ln\frac{yu^*}{\upsilon} + B \qquad \text{Overlap} - \text{layer} \tag{2.21}$$

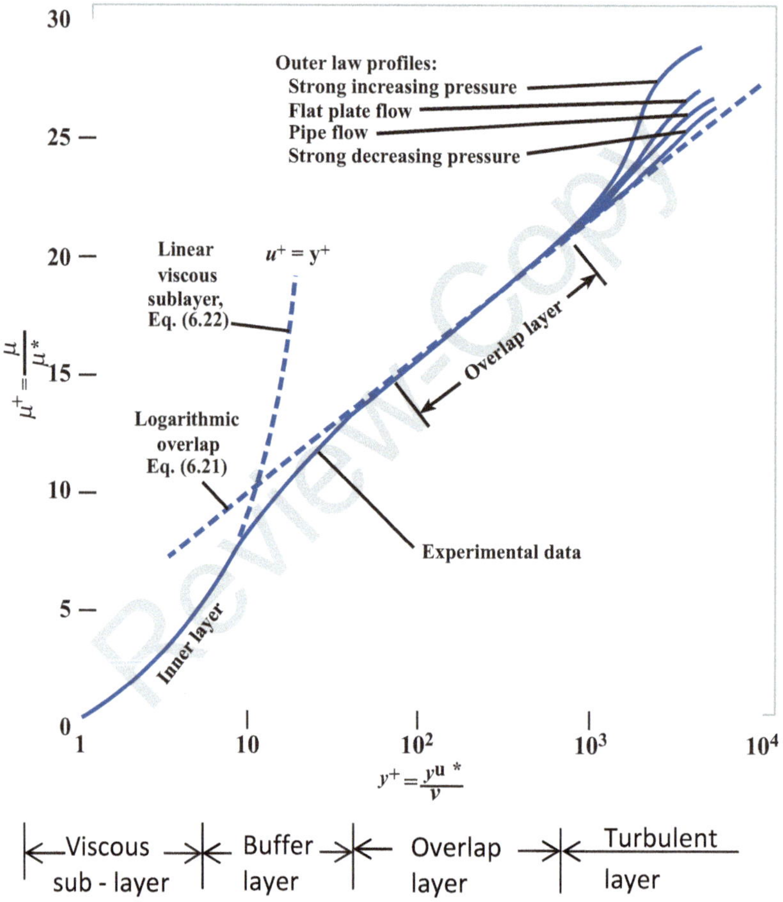

Fig. (2.10). Experimental verification of the inner - outer - and overlap - layer laws relating velocity profiles in turbulent wall flow [5].

Over the full range of turbulent wall flows the dimensionless constant k and B are found to have the approximate value $k = 0.41$ and $B = 5$. eq. (2.21) is called the logarithmic – overlap layer law.

Thus by dimensional reasoning and physical in sight we in far that a plot of u versus $\ln y$ in a turbulent – shear layer will show a cured wall region. A curved outer region, and a straight – line logarithmic overlap. Fig. (**2.10**) shows that this is exactly the case. The four outer – law profiles shown all merge smoothly with pressure gradient. The wall law is unique and follows the liner viscose the logarithmic – overlap – law but have different magnitudes because they vary in external relation.

$$u^+ = \frac{u}{u^*} = \frac{yu^*}{v} = y^+ \quad \textbf{(2.22)}$$

From the wall to about $y^+ = 5$, there after curving over to merge with the logarithmic law at a bout $y^+ = 30$.

Thus we can use eq. (2.22) as an excellent approximation to solve nearly every turbulent – flow problem presented.

2.7. TURBULENT VELOCITY PROFILE

Unlike laminar flow, the expressions for velocity profile in turbulent flow are based on both analysis and measurements, and thus they are semi – experimental data.

Typical velocity profiles for fully developed laminar and turbulent given in Fig. (**2.11**).

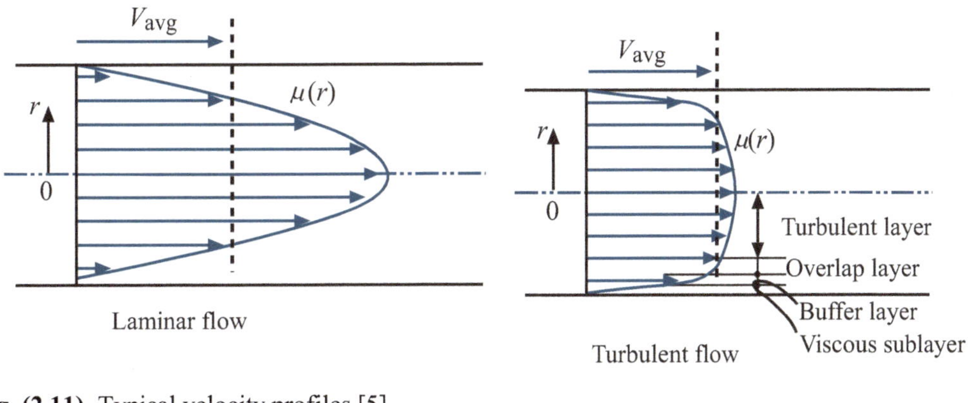

Fig. (2.11). Typical velocity profiles [5].

2.8. TURBULENT FLOW SOLUTION

2.8.1. Flow in Pipes

For turbulent pipe flow we need not solve a differential equation as in laminar flow, but instead proceed with the logarithmic law. Assume that eq. (2.21) correlates the local mean velocity $u_{(r)}$ all the way across the pipe [1, 6].

$$\frac{u_r}{u^*} = \frac{1}{k}\ln\frac{(R-r)u^*}{\upsilon} + B \tag{2.23}$$

Where we have replaced y by $(R-r)$. Computing the average velocity from this profile.

$$V = \frac{Q}{A} = \frac{1}{\pi R^2}\int_0^R u^*\left(\frac{1}{k}\ln\frac{(R-r)u^*}{\upsilon} + B\right)2\pi r\,dr$$

$$V = \frac{1}{2}u^*\left(\frac{2}{k}\ln\frac{Ru^*}{\upsilon} + 2B - \frac{3}{k}\right) \tag{2.23a}$$

Introducing $k = 0.41$ and $B = 5.0$ we obtain numerically

$$\frac{V}{u^*} = 2.44\ln\frac{Ru^*}{\upsilon} + 1.34 \tag{2.24}$$

This looks only marginally interesting until we realize that $\frac{V}{u^*}$ is directly related to

$$\frac{V}{u^*} = \left(\frac{\rho V}{\tau_w}\right)^{\frac{1}{2}} = \left(\frac{8}{f}\right)^{\frac{1}{2}} \tag{2.25}$$

Moreover, the argument of the logarithm in eq. (2.26) is equivalent to

$$\frac{Ru^*}{\upsilon} = \frac{\frac{1}{2}Vdu^*}{\upsilon\, V} = \frac{1}{2}Re_d\left(\frac{8}{f}\right)^{\frac{1}{2}} \tag{2.26}$$

Substituting eq., (2.27), in (2.26) we get

$$\frac{V}{u^*} = 2.44 \ln Re_d (f)^{\frac{1}{2}} - 2.89$$

Or,

$$\frac{1}{\sqrt{f}} = 0.863 \ \ln Re_d (f)^{\frac{1}{2}} - 1.02$$

$$\frac{1}{\sqrt{f}} = 1.99 \ \log Re_d (f)^{\frac{1}{2}} - 1.02 \qquad \textbf{(2.27)}$$

In other words, by simply computing the mean velocity from logarithmic –law correlation.

We obtain the relation between friction factor and Reynolds number for turbulent pipe flow.

Thus accepted formula for a smooth – walled pipe. There are many alternate approximation in the literature from which f can be computed explicitly from Re_d.

$$f = \begin{cases} 0.316 \ Re_d^{-\frac{1}{4}} & 4000 < Re < 10^5 \quad H.Blasius \ 1911 \\ \left(1.8 \ \log \dfrac{Re_d}{6.1}\right)^{-2} & C.F.Colebrook \ 1938 - 1939 \end{cases}$$

Therefore the pressure drop across a horizontal smooth pipe:

$$h_f = \frac{\Delta P}{\rho g} = f \frac{LV^2}{d2g} = 0.316 \left(\frac{\mu}{\rho V d}\right)^{\frac{1}{4}} . \frac{L}{d} . \frac{V^2}{2g}$$

Or,

$$\Delta P = 0.158 \ \rho^{\frac{3}{4}} \ \mu^{\frac{1}{4}} \ d^{-\frac{5}{4}} \ V^{\frac{7}{4}} \qquad \textbf{(2.28)}$$

Also,

$$\text{for } Q = AV = \frac{1}{4} \pi \ d^2 \ V$$

$$\therefore \quad \Delta P = 0.241\, L\, \rho^{\frac{3}{4}}\, \mu^{\frac{1}{4}}\, d^{-4.75}\, Q^{1.75} \qquad \textbf{(2.29)}$$

For a given flow rate Q, the turbulent pressure decreases with diameter even sharply than laminar flow. Thus, the quickest way to reduce required pumping pressure is to increase the pipe size, although of course the largest pipe more expensive.

Doubling the pipe size decreases ΔP by a factor of about 27 for given Q.

The maximum velocity in turbulent pipe flow is given by evaluating eq. (2.23) at $r = 0$

$$u_{(r)} = u_{max} \qquad \text{at} \qquad r = 0$$

$$\therefore \quad \frac{u_{max}}{u^*} = \frac{1}{k} \ln \frac{Ru^*}{\upsilon} + B \qquad \textbf{(2.30)}$$

Then combining this with equation (2.25) we get,

$$\frac{\mathrm{V}}{u_{max}} = \left(1 + 1.33\sqrt{f}\right)^{-1} \qquad \textbf{(2.31)}$$

The ratio varies with Re. No. and is Mach larger than the value 0.5

$$\mathrm{V} = \frac{1}{2}\, u_{max} \qquad \text{laminar flow.}$$

The turbulent velocity profile is very flat in the center and drop sharply to zero at the wall V = 0.85 u_{max} turbulent flow

2.8.2. Flow Between Parallel Plates

For turbulent flow between parallel plates a gain use the logarithm low [1,6], using the wall coordinate Y as shown in Fig. (**2.12**).

$$\frac{u_{(y)}}{u^*} = \frac{1}{k} \ln \frac{Yu^*}{\upsilon} + B \qquad 0 < Y < \frac{h}{2}$$

This distribution looks very much like the flat turbulent profile for pipe flow and the mean velocity is,

$$\mathrm{V} = \frac{2}{h}\int_0^{h/2} u\, dY = u^* \left(\frac{1}{k}\ln\frac{hu^*}{2\upsilon} + B - \frac{1}{k}\right) \quad \textbf{(2.32)}$$

recalling for $\frac{\mathrm{V}}{u^*} = \left(\frac{8}{f}\right)^{\frac{1}{2}}$ we find that

$$\frac{1}{\sqrt{f}} = 2.0\ \log\left(Re_{D_h}\sqrt{f}\right) - 1.19 \quad \textbf{(2.33)}$$

For flow in ducts $D_h = \frac{4A}{P} = \frac{cross - sectionarea}{wetted\ perimeter}$

$\therefore\ D_h = \frac{4\ bh}{2b + 2h}$ hence for parallel plate and unit depth $i.e. b = 1$

$\therefore\ D_h = 2h$

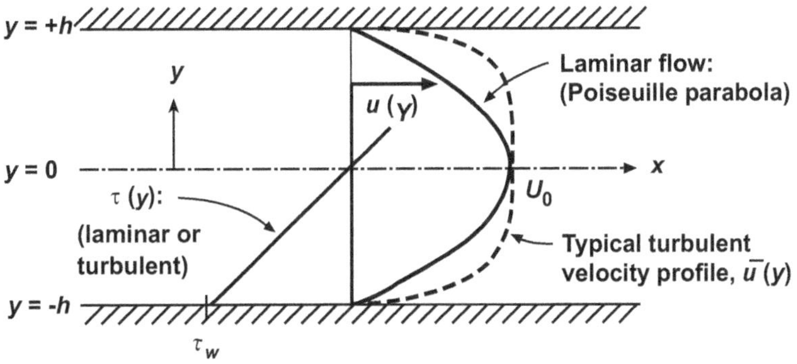

Fig. (2.12). Fully developed laminar and turbulent flow in a channel [7].

2.9. EFFECT OF ROUGH WALLS

The turbulent flow is strongly affected by roughness. In Fig. (**2.13**) the linear viscous sublayer only extends out to $y^+ = y\,u^*/\upsilon = 5$. Thus, compored with the diameter the sublayer thickness y_s is only

$$\frac{y_s}{d} = \frac{5\upsilon/u^*}{d} \quad \textbf{(2.34)}$$

and from eq. (2.34)

$$\frac{y_s}{d} = \frac{5\upsilon/u^*}{d}\cdot\frac{V}{V} = \frac{14.1}{Re_d\, f^{\frac{1}{2}}} \quad \textbf{(2.35)}$$

For example at $Re_d = 10^5, f = 0.018, \frac{y_s}{d} = 0.001$.

A wall roughness of about $0.001d$ will break up the sublayer and profoundly change the wall in Fig. (**2.13**).

Measurement of $u_{(y)}$ in turbulent rough – wall flow by Nikuradse show as in Fig (**2.14a**), that a roughness height $\in$ will force the logarithm – low profile outward on the abscissa by amount approximately equal to ln $\in^+$, where $\in^+ = \frac{\in u^*}{\upsilon}$. The slope of the logarithm law remains the same, $\frac{1}{k}$ but shift outward causes the constant B to be less by an amount [3, 4].

$\Delta B = \frac{1}{k}\ln \in^+$ The actual measured values ΔB are shown in Fig. (**2.14b**) for sand – grain roughness and commercially rough pipes.

Fig. (**2.5b**) reveals three regimes of rough wall

- $\in u^+/\upsilon < 5$: Hydraulically smooth walls, no effect of roughness on friction.
- $5 < \in u^+/\upsilon < 70$: Transitional roughness, moderate Reynolds – number effect.
- $\in u^+/\upsilon > 70$: Fully rough flows sublayer totally broken up and friction independent of Reynolds number.

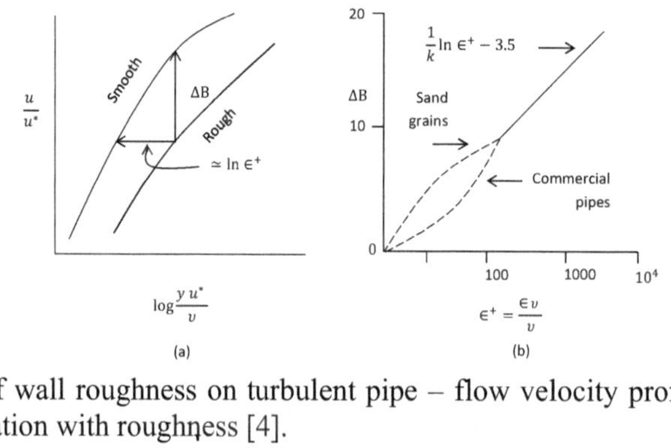

Fig. (2.14). Effect of wall roughness on turbulent pipe – flow velocity profile. (**a**) logarithm-law downshift (**b**) correlation with roughness [4].

For fully rough flow, $\epsilon^{+} > 70$, the data in Fig. (**2.14b**) follow the straight line

$$\Delta B = \frac{1}{k} \ln \epsilon^{+} - 3.5 \quad \textbf{(2.36)}$$

And the logarithm law modified for roughness becomes,

$$u^{+} = \frac{1}{k} \ln y^{+} + B - \Delta B = \frac{1}{k} \ln \frac{y}{\epsilon} + 8.5 \quad \textbf{(2.37)}$$

The viscosity vanishes and hence fully rough flow is independent of Reynolds number. Equation (2.32) could be integrated to obtain the average velocity in the pipe.

$$\frac{\mathrm{v}}{u^{*}} = 2.44 \ln \frac{d}{\epsilon} + 3.2 \quad \textbf{(2.38)}$$

Or

$$\frac{1}{\sqrt{f}} = -2.0 \log \frac{\epsilon/d}{3.7} \quad \text{fully rough wall} \quad \textbf{(2.39)}$$

There are no Reynolds effects, hence the head loss varies exactly as the square of the velocity in this case.

2.10. DERIVATION OF REYNOLDS STRESS EQUATION OF MOTION FOR TURBULENT FLOW

The equations of motion for real can be developed from consideration of the forces acting on a small element of the fluid, including the shear stresses generated by fluid motion and viscosity. The derivation of these equations, called the Navier – stokes eq. of motion for as incompressible fluid are [1, 8]:

$$\frac{\partial u}{\partial t} + u\frac{\partial u}{\partial x} + \mathrm{v}\frac{\partial u}{\partial y} + w\frac{\partial u}{\partial z} = -\frac{1}{\rho}\frac{\partial P}{\partial x} + \upsilon\left[\frac{\partial^2 u}{\partial x^2} + \frac{\partial^2 u}{\partial y^2} + \frac{\partial^2 u}{\partial z^2}\right] \quad \textbf{(2.40)}$$

$$\frac{\partial \mathrm{v}}{\partial t} + u\frac{\partial \mathrm{v}}{\partial x} + \mathrm{v}\frac{\partial \mathrm{v}}{\partial y} + w\frac{\partial \mathrm{v}}{\partial z} = -\frac{1}{\rho}\frac{\partial P}{\partial y} + \upsilon\left[\frac{\partial^2 \mathrm{v}}{\partial x^2} + \frac{\partial^2 \mathrm{v}}{\partial y^2} + \frac{\partial^2 \mathrm{v}}{\partial z^2}\right] \quad \textbf{(2.41)}$$

$$\frac{\partial \mathrm{w}}{\partial t} + u\frac{\partial \mathrm{w}}{\partial x} + \mathrm{v}\frac{\partial \mathrm{w}}{\partial y} + w\frac{\partial \mathrm{w}}{\partial z} = -\frac{1}{\rho}\frac{\partial P}{\partial z} + \upsilon\left[\frac{\partial^2 \mathrm{w}}{\partial x^2} + \frac{\partial^2 \mathrm{w}}{\partial y^2} + \frac{\partial^2 \mathrm{w}}{\partial z^2}\right] \quad \textbf{(2.42)}$$

$$\text{Continuity eq.} \quad \frac{\partial u}{\partial x}+\frac{\partial \mathrm{v}}{\partial y}+\frac{\partial \mathrm{w}}{\partial z}=0 \qquad \textbf{(2.43)}$$

x, y, z Cartesian coordination

u, v, w the corresponding velocity comp.

ρ, P, υ and $t,$ density, power, kinematic viscosity, time

This eq.s' is valid for turbulent flow, when we substituting the corresponding velocities as follows.

$$u=\bar{u}+u' \;,\quad \mathrm{v}=\bar{\mathrm{v}}+\mathrm{v}' \;,\quad w=\bar{w}+w' \;,\quad P=\bar{P}+P' \qquad \textbf{(2.44)}$$

Eq. (4) becomes

$$\frac{\partial \bar{u}}{\partial x}+\frac{\partial u'}{\partial x}+\frac{\partial \bar{\mathrm{v}}}{\partial y}+\frac{\partial \mathrm{v}'}{\partial y}+\frac{\partial \bar{w}}{\partial z}+\frac{\partial w'}{\partial z}=0 \qquad \textbf{(2.45)}$$

Take the mean value of this eq.

$$\frac{\partial \bar{u}}{\partial x}+\frac{\partial \bar{\mathrm{v}}}{\partial y}+\frac{\partial \bar{w}}{\partial z}=0 \qquad \textbf{(2.46)}$$

Since $\frac{\partial u'}{\partial x}=\frac{\partial}{\partial x}(\overline{u'})=0\; etc.$ then substituting (2.50) from (2.49) gives

$$\frac{\partial u'}{\partial x}+\frac{\partial \mathrm{v}'}{\partial y}+\frac{\partial w'}{\partial z}=0 \qquad \textbf{(2.47)}$$

Eqs. (2.46), (2.47) show that the mean velocity comp. and fluctuating velocity comp. respectively satisfy the continuity equations.

Substituting equation (2.44) into x – momentum eq. (2.40) then,

$$\frac{\partial \bar{u}}{\partial t}+\frac{\partial u'}{\partial t}+(\bar{u}+u')\frac{\partial}{\partial x}(\bar{u}+u')+(\bar{\mathrm{v}}+\mathrm{v}')\frac{\partial}{\partial y}(\bar{u}+u')$$
$$+(\bar{\mathrm{w}}+\mathrm{w}')\frac{\partial}{\partial z}(\bar{u}+u')$$

$$= -\frac{1}{\rho}\frac{\partial \bar{P}}{\partial x} - \frac{1}{\rho}\frac{\partial P'}{\partial x} + \upsilon\left[\frac{\partial^2}{\partial x^2} + \frac{\partial^2}{\partial y^2} + \frac{\partial^2}{\partial z^2}\right](\bar{u} + u')$$

Take the mean value of this eq. and assume steady mean flow then.

$$\bar{u}\frac{\partial \bar{u}}{\partial x} + \bar{v}\frac{\partial \bar{u}}{\partial y} + \bar{w}\frac{\partial \bar{u}}{\partial z} + \underbrace{u'\frac{\partial u'}{\partial x} + v'\frac{\partial u'}{\partial y} + w'\frac{\partial u'}{\partial z}}_{(a)}$$

$$= -\frac{1}{\rho}\frac{\partial \bar{P}}{\partial x} + \upsilon\left[\frac{\partial^2 \bar{u}}{\partial x^2} + \frac{\partial^2 \bar{u}}{\partial y^2} + \frac{\partial^2 \bar{u}}{\partial z^2}\right] \quad \textbf{(2.48)}$$

In obtaining the above eq. (2.49), we have used the following assumptions.

a) $\frac{\partial \bar{u}}{\partial t} = 0$ since the main flow is steady

b) $u'\frac{\partial u'}{\partial t} = 0$ since $\frac{\partial u'}{\partial t} = \lim_{T_0\to\infty}\frac{1}{T_0}\int_0^{T_0}\frac{\partial u'}{\partial t}dt = \lim_{T_0\to\infty}\frac{u'_{T_0} - u'_{(0)}}{T_0} = 0$

c) $u'\frac{\partial \bar{u}}{\partial t} = \overline{u'}\frac{\partial \bar{u}}{\partial t} = 0$ since $\overline{u'} = 0$ fluctuation vanish

d) $\bar{u} = \frac{\partial u'}{\partial t} = \bar{u}\frac{\partial \overline{u'}}{\partial t} = \bar{u}\frac{\partial}{\partial x}(\overline{u'}) = 0$ since $\overline{u'} = 0$

e) $v'\frac{\partial \bar{u}}{\partial y} = w'\frac{\partial \bar{u}}{\partial z} = \bar{v}\frac{\partial u'}{\partial y} = \bar{w}\frac{\partial u'}{\partial z} = 0$ similar arguments

f) $\frac{\partial^2 u'}{\partial x^2} = \frac{\partial^2}{\partial x^2}(\overline{u'}) = 0$ $\overline{u'} = 0$

g) Similarly $\frac{\partial^2 u'}{\partial y^2}, \frac{\partial^2 u'}{\partial z^2}$ ……

Now consider term (a) into eq. (2.49) we have

$$u'\frac{\partial u'}{\partial x} + v'\frac{\partial u'}{\partial y} + w'\frac{\partial u'}{\partial z}$$

$$= \frac{\partial}{\partial x}(u'^2) + \frac{\partial}{\partial y}(u'v') + \frac{\partial}{\partial z}(u'w') - u'\left(\frac{\partial u'}{\partial x} + \frac{\partial v'}{\partial y} + \frac{\partial w'}{\partial z}\right)$$

But from eq. (2.47) $\frac{\partial u'}{\partial x} + \frac{\partial v'}{\partial y} + \frac{\partial w'}{\partial z} = 0$ and so

$$u'\frac{\partial u'}{\partial x}+\mathrm{v}'\frac{\partial u'}{\partial y}+w'\frac{\partial u'}{\partial z}=\frac{\partial}{\partial x}(u'^2)+\frac{\partial}{\partial y}(u'\mathrm{v}')+\frac{\partial}{\partial z}(u'w')$$
$$=\frac{\partial}{\partial x}\left(\overline{u'^2}\right)+\frac{\partial}{\partial y}(\overline{u'\mathrm{v}'})+\frac{\partial}{\partial z}(\overline{u'w'})$$

Hence after multiplying by ρ, eq. (2.48) can be expressed in the form.

$$\rho\left(\bar{u}\,\frac{\partial\bar{u}}{\partial x}+\bar{\mathrm{v}}\frac{\partial\bar{u}}{\partial y}+\overline{w}\frac{\partial\bar{u}}{\partial z}\right)=-\frac{\partial\bar{P}}{\partial x}+\mu\left(\frac{\partial^2\bar{u}}{\partial x^2}+\frac{\partial^2\bar{u}}{\partial y^2}+\frac{\partial^2\bar{u}}{\partial z^2}\right)+\frac{\partial}{\partial x}\left(-\rho\,\overline{u'^2}\right)+\frac{\partial}{\partial y}(-\rho\,\overline{u'\mathrm{v}'})+\frac{\partial}{\partial z}(-\rho\,\overline{u'w'}) \quad \textbf{(2.49)}$$

It is now convenient to introduce the Reynolds stress (sometimes called apparent stress or eddy stress) defined as follow:

$$\tau_{xx}=-\rho\,\overline{u'^2},\qquad \tau_{yy}=-\rho\,\overline{\mathrm{v}'^2},\qquad \tau_{zz}=-\rho\,\overline{w'^2}$$
$$\tau_{xy}=\tau_{yx}=-\rho\,\overline{u'\mathrm{v}'},\ \ \tau_{yz}=\tau_{zy}=-\rho\,\overline{u'w'}$$
$$\tau_{zx}=\tau_{xz}=-\rho\,\overline{u'w'}$$

Then eq. (2.51) becomes

$$\rho\left(\bar{u}\,\frac{\partial\bar{u}}{\partial x}+\bar{\mathrm{v}}\frac{\partial\bar{u}}{\partial y}+\overline{w}\frac{\partial\bar{u}}{\partial z}\right)=-\frac{\partial\bar{P}}{\partial x}+\mu\nabla^2\bar{u}+\frac{\partial\tau_{xx}}{\partial x}+\frac{\partial\tau_{xy}}{\partial y}+\frac{\partial\tau_{xz}}{\partial z} \quad \textbf{(2.50)}$$

Similarly:

$$\rho\left(\bar{u}\,\frac{\partial\bar{\mathrm{v}}}{\partial x}+\bar{\mathrm{v}}\frac{\partial\bar{\mathrm{v}}}{\partial y}+\overline{w}\frac{\partial\bar{\mathrm{v}}}{\partial z}\right)=-\frac{\partial\bar{P}}{\partial y}+\mu\nabla^2\bar{\mathrm{v}}+\frac{\partial\tau_{yx}}{\partial x}+\frac{\partial\tau_{yy}}{\partial y}+\frac{\partial\tau_{yz}}{\partial z} \quad \textbf{(2.51)}$$

$$\rho\left(\bar{u}\,\frac{\partial\overline{w}}{\partial x}+\bar{\mathrm{v}}\frac{\partial\overline{w}}{\partial y}+\overline{w}\frac{\partial\overline{w}}{\partial z}\right)=-\frac{\partial\bar{P}}{\partial z}+\mu\nabla^2\overline{w}\frac{\partial\tau_{zx}}{\partial x}+\frac{\partial\tau_{zy}}{\partial y}+\frac{\partial\tau_{zz}}{\partial z} \quad \textbf{(2.52)}$$

Eqs. (2.52, 2.53, 2.54) are the Reynolds equations for turbulent flow.

Note that they are of the same form as the Navior–Stokes equations, apart from the addition of the Reynolds stress terms [13].

We may also introduce of the total stress terms defines as follows:

Normal Shear Turbulent

↓ ↓ ↓

$$P_{xx} = -\bar{P} + 2\mu\frac{\partial\bar{u}}{\partial x} + \tau_{xx}$$

$$P_{yy} = -\bar{P} + 2\mu\frac{\partial\bar{v}}{\partial x} + \tau_{yy}$$

$$P_{zz} = -\bar{P} + 2\mu\frac{\partial\bar{w}}{\partial x} + \tau_{zz}$$

$$P_{xy} = P_{yx} = \mu\left(\frac{\partial\bar{u}}{\partial y} + \frac{\partial\bar{v}}{\partial x}\right) + \tau_{xy}$$

$$P_{yz} = P_{zy} = \mu\left(\frac{\partial\bar{v}}{\partial z} + \frac{\partial\bar{w}}{\partial y}\right) + \tau_{yz}$$

$$P_{zx} = P_{xz} = \mu\left(\frac{\partial\bar{w}}{\partial x} + \frac{\partial\bar{u}}{\partial z}\right) + \tau_{zx}$$

The first column denotes stresses due to the mean pressure, the second column denotes stress due to viscosity and the third column denotes stress due to fluctuation.

The Reynolds equations may then be expressed in the form:

$$\rho\frac{D\bar{u}}{Dt} = \frac{\partial P_{xx}}{\partial x} + \frac{\partial P_{xy}}{\partial y} + \frac{\partial P_{xz}}{\partial z}$$
$$\rho\frac{D\bar{v}}{Dt} = \frac{\partial P_{yx}}{\partial x} + \frac{\partial P_{yy}}{\partial y} + \frac{\partial P_{yz}}{\partial z}$$
$$\rho\frac{D\bar{w}}{Dt} = \frac{\partial P_{zx}}{\partial x} + \frac{\partial P_{zy}}{\partial y} + \frac{\partial P_{zz}}{\partial z}$$

Verify these equations by substituting the valves of the total stress given in the above equations.

CHAPTER 3

The Dynamics of Turbulence

Abstract: Two major questions arise. First, how is the kinetic energy of turbulence maintained? Second, why are vorticity and vortex stretching so important to the study of turbulence? To help answer these questions. First, derive equations for the kinetic energy of the mean flow and that of turbulence.

In this chapter, the derivation of two – dim. K.E. was presented in full details and then mads assumption to find out the solution of the equation in simple way. The approximated velocity profile with strong pressure gradient and at the separation point are presented.

Keywords: Separation point, TKE, Two dimensional, Velocity profile.

3.1. INTRODUCTION

In chapter two we studied the effect of turbulent velocity fluctuations on the flow. We now turn to the other side of the issue. The major questions arise, how is the kinetic energy of the turbulence maintained? First derive equations for the kinetic energy of the mean flow and that of the turbulence.

3.2. TURBULENT KINETIC ENERGY (TKE)

The transport equation of TKE describes how mean flow feeds kinetic energy into turbulence. The transport of TKE also plays a vital role in the development of turbulence models [1, 9].

The instantaneous kinetic energy $K(t)$ is the sum of mean kinetic $\bar{K}$ energy

$$\bar{K} = (\bar{U}^2 + \bar{V}^2 + \bar{W}^2)/2 \quad \textbf{(3.1)}$$

And turbulent kinetic energy

$$k = \left(u'^2 + \mathrm{v}'^2 + w'^2\right)/2 \quad \textbf{(3.2)}$$

$$i.e.\ \ K(t) = \bar{K} + k \quad \textbf{(3.3)}$$

Jafar Mehdi Hassan, Riyadh S. Al-Turaihi, Salman Hussien Omran, Laith Jaafer Habeeb, Alamaslamani Ammar Fadhil Shnawa

assuming that the fluid is incompressible, we use Navier – Stokes equations, equations (2.40) to (2.43).

For two dimensional boundary layer approximation and steady state equation (2.41) becomes:

$$(\bar{U}+u')\frac{\partial(\bar{U}+u')}{\partial x}+(\bar{V}+u')\frac{\partial(\bar{U}+u')}{\partial y}=-\frac{1}{\rho}\frac{\partial(\bar{P}+P')}{\partial x}+\upsilon\left(\frac{\partial^2(\bar{U}+u')}{\partial x^2}+\frac{\partial^2(\bar{U}+u')}{\partial y^2}\right) \quad \textbf{(3.4)}$$

Multiply each of the component of equation (3.4) by its respective fluctuation velocity, u' yields:

$$u'(\bar{U}+u')\frac{\partial(\bar{U}+u')}{\partial x}+u'(\bar{V}+u')\frac{\partial(\bar{U}+u')}{\partial y}=-\frac{1}{\rho}u'\frac{\partial(\bar{P}+P')}{\partial x}+\upsilon u'\left(\frac{\partial^2(\bar{U}+u')}{\partial x^2}+\frac{\partial^2(\bar{U}+u')}{\partial y^2}\right) \quad \textbf{(3.5)}$$

The equation (3.5) can be further arranged to:

$$\bar{U}u'\frac{\partial\bar{U}}{\partial x}+\bar{U}u'\frac{\partial u'}{\partial x}+u'^2\frac{\partial\bar{U}}{\partial x}+u'^2\frac{\partial u'}{\partial x}+\bar{V}u'\frac{\partial\bar{U}}{\partial y}+\bar{V}u'\frac{\partial u'}{\partial y}+u'\mathrm{v}'\frac{\partial\bar{U}}{\partial y}+u'\mathrm{v}'\frac{\partial u'}{\partial y}=-\frac{1}{\rho}u'\frac{\partial\bar{P}}{\partial x}-\frac{1}{\rho}u'\frac{\partial P'}{\partial x}+\upsilon u'\left(\frac{\partial^2\bar{U}}{\partial x^2}+\frac{\partial^2\bar{U}}{\partial y^2}\right)+\upsilon u'\left(\frac{\partial^2 u'}{\partial x^2}+\frac{\partial^2 u'}{\partial y^2}\right) \quad \textbf{(3.6)}$$

Taking the time average of equation (3.6),

$$\overline{\bar{\phi}\phi'}=\overline{\phi\phi'}=0 \qquad \overline{\overline{\phi\varphi}}=\bar{\phi}\,\bar{\varphi} \qquad \overline{\bar{\phi}+\bar{\varphi}}=\bar{\phi}+\bar{\varphi} \quad \textbf{(3.7)}$$

We obtain,

$$\underbrace{\overline{\bar{U}u'\frac{\partial\bar{U}}{\partial x}}}_{=0}+\overline{\bar{U}u'\frac{\partial u'}{\partial x}}+\overline{u'^2\frac{\partial\bar{U}}{\partial x}}+\overline{u'^2\frac{\partial u'}{\partial x}}+\underbrace{\overline{\bar{V}u'\frac{\partial\bar{U}}{\partial y}}}_{=0}+\overline{\bar{V}u'\frac{\partial u'}{\partial y}}+\overline{u'\mathrm{v}'\frac{\partial\bar{U}}{\partial y}}+\overline{u'\mathrm{v}'\frac{\partial u'}{\partial y}}$$

$$=\underbrace{-\overline{\frac{1}{\rho}u'\frac{\partial\bar{P}}{\partial x}}}_{0=}-\overline{\frac{1}{\rho}u'\frac{\partial P'}{\partial x}}+\underbrace{\overline{\upsilon u'\left[\frac{\partial^2\bar{U}}{\partial x^2}+\frac{\partial^2 U}{\partial y^2}\right]}}_{=0}+\overline{\upsilon u'\left[\frac{\partial^2 u'}{\partial x^2}+\frac{\partial^2 u'}{\partial y^2}\right]} \quad \textbf{(3.8)}$$

The equation (3.8) further simplified to

$$\overline{U u' \frac{\partial u'}{\partial x}} + \overline{u'}^2 \frac{\partial \bar{U}}{\partial x} + \overline{u'^2 \frac{\partial u'}{\partial x}} + \overline{\bar{V} u' \frac{\partial u'}{\partial y}} + \overline{u' v'} \frac{\partial \bar{U}}{\partial y} + \overline{u' v' \frac{\partial u'}{\partial y}} = -\frac{1}{\rho} u' \frac{\partial P'}{\partial x} + \upsilon u' \left(\frac{\partial^2 u'}{\partial x^2} + \frac{\partial^2 u'}{\partial y^2} \right) \quad \textbf{(3.9)}$$

Using the property:

$$u \frac{\partial u}{\partial x} = \frac{1}{2} \frac{\partial u^2}{\partial x} \quad \textbf{(3.10)}$$

The equation (3.9) can be further re – arranged:

$$\bar{U} \frac{1}{2} \frac{\partial \overline{u'^2}}{\partial x} + \bar{V} \frac{1}{2} \frac{\partial u'^2}{\partial y} + \overline{u'^2} \frac{\partial \bar{U}}{\partial x} + \overline{\bar{V} u'} \frac{\partial \bar{U}}{\partial y} + u'^2 \frac{\partial u'}{\partial x} + u' v' \frac{\partial u'}{\partial y} = -\frac{1}{\rho} u' \frac{\partial P'}{\partial x} + \upsilon u' \left(\frac{\partial^2 u'}{\partial x^2} + \frac{\partial^2 u'}{\partial y^2} \right) \quad \textbf{(3.11)}$$

By multiplying v' with equation (2.42) and performing time averaging, the following equation can be obtained

$$\bar{U} \frac{1}{2} \frac{\partial \overline{v'^2}}{\partial x} + \bar{V} \frac{1}{2} \frac{\partial \overline{v'^2}}{\partial y} + \overline{v'^2} \frac{\partial \bar{V}}{\partial x} + \overline{u' v'} \frac{\partial \bar{V}}{\partial x} + \overline{v'^2 \frac{\partial v'}{\partial y}} + \overline{u' v' \frac{\partial u'}{\partial x}} = -\frac{1}{\rho} v' \frac{\partial P'}{\partial y} + \upsilon v' \nabla^2 v' \quad \textbf{(3.12)}$$

Adding equation (3.11) and (3.12) and defining kinetic energy with velocity fluctuations

$$k = \frac{1}{2} \left(u'^2 + v'^2 \right) \quad \textbf{(3.13)}$$

We obtained

$$\bar{U} \frac{\partial k}{\partial x} + \bar{V} \frac{\partial k}{\partial y} - (-\overline{u' v'}) \frac{\partial \bar{U}}{\partial y} - (-\overline{u' v'}) \frac{\partial \bar{V}}{\partial x} + \frac{\partial}{\partial y} \left[v' \left(k + \frac{P}{\rho} \right) \right] + \frac{\partial}{\partial x} \left[\overline{u' \left(k + \frac{P'}{\rho} \right)} \right] + \left[\overline{u'^2} \frac{\partial \bar{U}}{\partial x} + \overline{v'^2} \frac{\partial \bar{v}}{\partial y} \right] - \upsilon [u' \nabla^2 u' + v' \nabla^2 v'] = 0 \quad \textbf{(3.14)}$$

Where

1- $\epsilon_d = -\upsilon[u'\nabla^2 u' + v'\nabla^2 v']$

= Dissipation of TKE by molecular viscosity force.

2- From the continuity equation, we reasoned that there must be a correlation between u' and v', therefore $u' \sim v'$ and $\overline{u'v'} \approx u^{*2} = \frac{\tau}{\rho}$ the term could be

$$u^{*2}\left[\frac{\partial U}{\partial x} + \frac{\partial v}{\partial y}\right] \approx 0$$

Also the term represents the effect of fluctuation of Reynolds stress about the mean value and in shear flow, it is usually much smaller that velocity shear $\frac{\partial \bar{U}}{\partial y}$.

3- also the normal stresses along the *x*- direction normally neglected. Then $\frac{\partial}{\partial x}\left[u'\left(k + \frac{P'}{\rho}\right)\right] \approx 0$.

This yields the following final results which are approximately as derived by Townsend book of "the structure of turbulent shear flow" [9]:

$$\bar{U}\frac{\partial k}{\partial x} + \bar{V}\frac{\partial k}{\partial y} - (-u'v')\frac{\partial \bar{U}}{\partial y} + \frac{\partial}{\partial y}\left[v'\left(\frac{P'}{\rho} + k\right)\right] + \epsilon_d = 0 \tag{3.15}$$

Advection Production Diffusion Dissipation

ADV = transport of TKE by mean flow.
PRO = generation of TKE by mean flow velocity shear $\partial\bar{U}/\partial y$.
DIFF = redistribution of TKE by turbulent itself.
DISS = conversion of TKE into heat.

3.3 SOLUTION OF TKE

This equation contains many new unknowns so it cannot be useful employed for the basic closure problem. If can however be used for experimental B.L. observation [1, 3].

A typical case is approximately, in particular close to the wall, (Fig. **3.1**).

PROD ≈ DISS

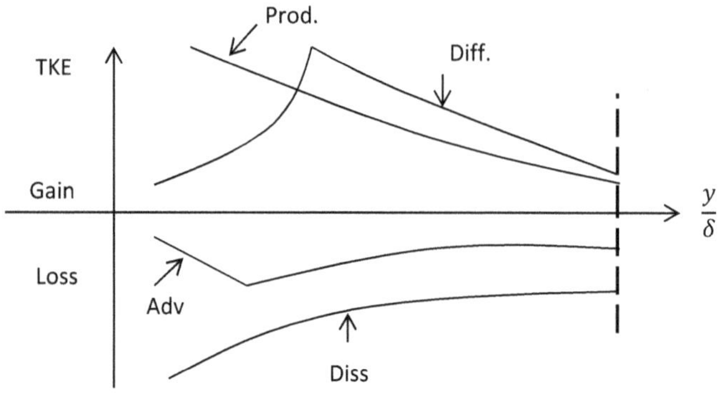

Fig. (3.1). Boundary layers close to the wall.

This feature may be applied in approximate exactly in inner regime (IR).

$$\epsilon_d \approx -\overline{u'v'}\ \frac{\partial \bar{U}}{\partial y} \quad \textbf{(3.16)}$$

To find approximate simplified for Diss., Prod. And Diff., Adv., is to be ignored altogether as it is small regardless.

a) Dissipation:- Consider a simplified turbulent eddy of size λ and using a fluctuating velocity u'. Assume it have a shape of ball.

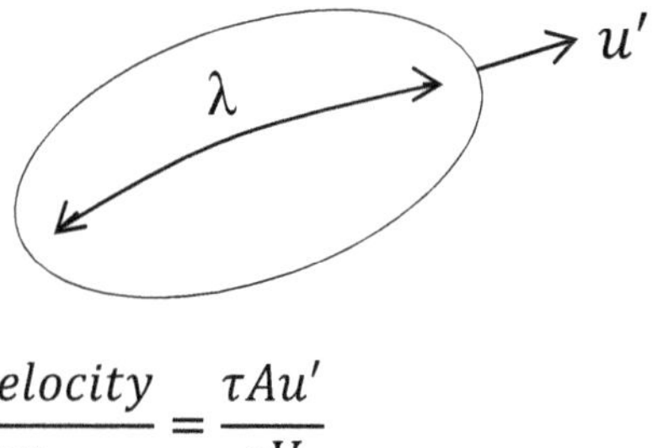

$$\epsilon_d = \frac{Drag \cdot velocity}{mass} = \frac{\tau A u'}{\rho V}$$

$$= \frac{\left(C_D \rho \frac{u'^2}{2}\right)\left(\frac{\pi}{4}\lambda^2\right) u'}{\rho \cdot \frac{\pi}{6}\lambda^3} = C_D \frac{3}{4}\frac{u'^3}{\lambda}$$

$$\epsilon_d = constant\ \frac{u'^3}{\lambda} \tag{3.17}$$

Atypical velocity scale for a fluctuating turbulent eddy.

$$u' = \sqrt{\bar{k}}\ \text{ then }\ \epsilon_d = C_1\,\bar{k}^{\frac{3}{2}}/\lambda \tag{3.17a}$$

b) Production: - From many boundary layer measurements it is known that

$$\left(\frac{-\overline{u'v'}}{\bar{k}}\right) \approx a_1 = constant \qquad 0.12 < a_1 < 0.18 \quad \text{usually} \quad a_1 \approx 0.1$$

$$\text{Then}\quad PROD = a_1\bar{k}\frac{\partial\bar{U}}{\partial y} \tag{3.18}$$

c) Diffusion: - Is mainly due to the turbulent agitation only weakly dependent on pressure fluctuations.

Additionally it is known that the main agent of diffusion is v', then:

$v'\left(\bar{k} + \frac{P'}{\rho}\right) \approx bv'^3$ neglecting pressure fluctuation and $\bar{k} \approx v'^2$

$$\text{or}\quad v'\left(\bar{k} + \frac{P'}{\rho}\right) \approx b\bar{k}^{\frac{3}{2}} \tag{3.19}$$

Because of the PROD ≈ DISS equality the IR is often call the equilibrium region in which in the absence of strong pressure gradient, (Fig. **3.2**):

$$-\overline{u'v'} = u^{*2}$$

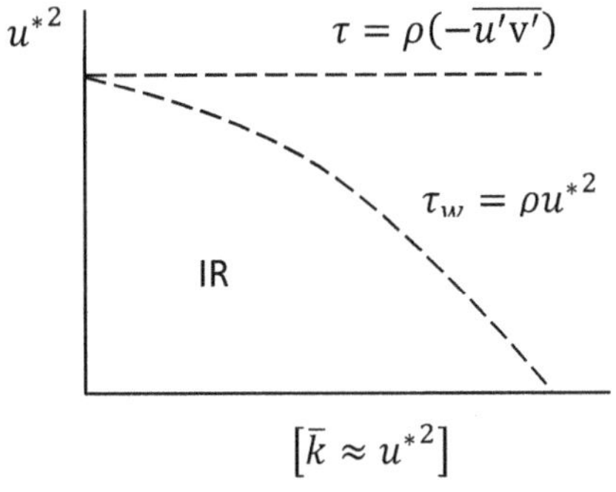

Fig. (3.2). Diffusion.

or $\bar{k}$ in IR approximately u^{*2}

since u^{*} becomes a measure

of fluctuating velocities

then $-\overline{u'v'}\,\frac{\partial \bar{U}}{\partial y} = C_1\,\bar{k}^{\frac{3}{2}}/\lambda$ or $C_1\frac{u^{*3}}{\lambda}$

Production $\approx$ Dissipation

If $\lambda \approx ky$ (proportional to the distance from the wall and k – Vonkarman constant) Then

$u^{*2}\frac{\partial \bar{U}}{\partial y} = C_1\frac{u^{*3}}{\lambda} = C_1\frac{u^{*3}}{ky}$ the integrating of both side with respect to y for $C_1 \approx 1$

$$\frac{\bar{U}}{u^{*}} = \frac{1}{k}\ln y + B \quad \text{Law of the wall.} \tag{3.20}$$

It is not always possible to assume that the shear stress near the wall hardly varies from the wall value τ_w. This would be unveiled in T.B.L. with strong pressure gradient. However, near the wall the T.B.L.'s velocity gradient becomes:
From momentum equation, the inertia terms are negligible then:

$$\underbrace{\bar{U}\frac{\partial \bar{U}}{\partial x} + \bar{V}\frac{\partial \bar{U}}{\partial y}}_{=0} = -\frac{1}{\rho}\frac{\partial P_\infty}{\partial x} + \frac{1}{\rho}\frac{\partial \tau}{\partial y} \tag{3.21}$$

Then

$$\frac{\partial \tau}{\partial y} \approx \frac{\partial P_\infty}{\partial x} \tag{3.22}$$

Integrating with respect to y from $y = 0$ to any region

$$\text{at} \begin{matrix} y=0 \\ y=\infty \end{matrix} \quad \left.\begin{matrix} \tau=\tau_w \\ \tau=\tau \end{matrix}\right\} \tag{3.23}$$

$$\therefore \tau = \tau_w + \frac{\partial P_\infty}{\partial x} y = \tau_w + \alpha y \qquad \textbf{(3.24)}$$

An improved version of the wall can be obtained from eq.(3.14), using equations (3.17,3.18,3.19) and omitting ADV term using the fluctuating velocity scale $u^* = \sqrt{\bar{k}} = \sqrt{\frac{\tau}{\rho}}$

Then

$$\frac{\tau}{\rho}\frac{\partial \bar{U}}{\partial x} = \frac{1}{ky}\left(\frac{\tau}{\rho}\right)^{3/2} + \frac{\partial}{\partial y}\left[b_1\left(\frac{\tau}{\rho}\right)^{3/2}\right]$$

Production dissipation diffusion

Let $\bar{y} = \frac{1}{ky}$ and dividing both side by $\frac{\tau}{\rho}$ then

$$\frac{\partial \bar{U}}{\partial y} = \bar{y}\left(\frac{\tau}{\rho}\right)^{1/2} + \frac{1}{\tau/\rho}\frac{\partial}{\partial y}\left[b_1\left(\frac{\tau}{\rho}\right)^{3/2}\right]$$

Substituting for $\tau = \tau_w + \alpha y$:

$$\frac{\partial \bar{U}}{\partial y} = \bar{y}\left(\frac{\tau_w + \alpha y}{\rho}\right)^{1/2} + \frac{1}{\left(\frac{\tau_w + \alpha y}{\rho}\right)}\frac{\partial}{\partial y}\left[b_1\left(\frac{\tau_w + \alpha y}{\rho}\right)^{3/2}\right]$$

$$= \bar{y}\left(\frac{\tau_w}{\rho}\right)^{1/2}\left(1 + \frac{\alpha y}{\tau_w}\right)^{1/2} + \frac{1}{\frac{\tau_w}{\rho}\left(1 + \frac{\alpha y}{\tau_w}\right)}\frac{\partial}{\partial y}\left[b_1\left(\frac{\tau_w}{\rho}\right)^{3/2}\left(1 + \frac{\alpha y}{\tau_w}\right)^{3/2}\right]$$

$$= \bar{y}\left(\frac{\tau_w}{\rho}\right)^{1/2}\left(1 + \frac{\alpha y}{\tau_w}\right)^{1/2} + \frac{b_1\left(\frac{\tau_w}{\rho}\right)^{3/2}}{\frac{\tau_w}{\rho}\left(1 + \frac{\alpha y}{\tau_w}\right)}\frac{3}{2}\left(1 + \frac{\alpha y}{\tau_w}\right)^{1/2} \times \frac{\alpha}{\tau_w}$$

Arranging the equation and substituting for u^*

$$\frac{\partial \bar{U}}{\partial y} = \frac{u^*}{ky}\left(1+\frac{\alpha y}{\tau_w}\right)^{1/2} + \frac{\frac{3}{2}u^* b_1 \frac{\alpha}{\tau_w}\left(1+\frac{\alpha y}{\tau_w}\right)^{1/2}}{\left(1+\frac{\alpha y}{\tau_w}\right)} \tag{3.25}$$

Or

$$\frac{\partial \bar{U}}{\partial y} = \frac{u^*}{ky}\sqrt{1+\frac{\alpha y}{\tau_w}}\left(1+\frac{3}{2}b_1 k \frac{\frac{\alpha y}{\tau_w}}{\left(1+\frac{\alpha y}{\tau_w}\right)}\right) \tag{3.26}$$

For $\frac{\alpha y}{\tau_w} \ll 1$ then

$$\frac{\partial \bar{U}}{\partial y} = \frac{u^*}{ky} \tag{3.27}$$

Integration eq. (3.25) with respect to y:

$$\int \partial \bar{U} = \frac{u^*}{ky}\int \frac{1}{y}\left(1+\frac{\alpha y}{\tau_w}\right)^{1/2} dy + \frac{3}{2}u^* b_1 \frac{\alpha}{\tau_w}\int \left(1+\frac{\alpha y}{\tau_w}\right)^{-1/2} dy$$

Or

$$\frac{\bar{U}}{u^*} = \underbrace{\frac{1}{k}\int \frac{1}{y}\left(1+\frac{\alpha y}{\tau_w}\right)^{1/2} dy}_{A} + \underbrace{\frac{3}{2}b_1 \frac{\alpha}{\tau_w}\int \left(1+\frac{\alpha y}{\tau_w}\right)^{-1/2} dy}_{B}$$

$$A = \frac{1}{k}\int \frac{1}{y}\left(1+\frac{\alpha y}{\tau_w}\right)^{\frac{1}{2}} dy$$

Let $\frac{\alpha}{\tau_w} = a$ and $(1+\alpha y)^{\frac{1}{2}} = u$

$$\therefore 1+\alpha y = u^2 \quad \text{or} \quad y = \frac{u^2-1}{a} \quad \text{and} \quad \partial y = \frac{2u}{a}du$$

$$\therefore \int \frac{1}{y}(1+\alpha y)^{\frac{1}{2}}\partial y = \int \frac{au}{u^2-1}\cdot\frac{2u}{a}du = \int \frac{2u^2}{u^2-1}du$$

Integration by partial fraction *i.e.*

$$\frac{2u^2}{u^2-1}=\frac{2u^2}{(u-1)(u+1)}=\frac{A}{(u-1)}+\frac{B}{(u+1)}$$
$$=2u^2=A(u+1)+B(u-1)$$
$$\text{Let } u=1 \rightarrow 2\times 1^2=2A+B\times 0 \rightarrow A=1$$
$$\text{Let } u=-1 \rightarrow 2\times(-1)^2=A\times 0+B\times(-2) \rightarrow B=-1$$

$$\therefore \int \frac{2u^2}{(u-1)(u+1)}du=\int\frac{1}{(u-1)}du-\int\frac{1}{(u+1)}du$$

$$\therefore \int \frac{1}{y}\left(1+\frac{\alpha y}{\tau_w}\right)^{\frac{1}{2}}\partial y=\ln\left[\frac{\sqrt{\left(1+\frac{\alpha y}{\tau_w}\right)}-1}{\sqrt{\left(1+\frac{\alpha y}{\tau_w}\right)}+1}\right]+c$$

$$\text{or } A=\frac{1}{k}\ln\left[\frac{\sqrt{\left(1+\frac{\alpha y}{\tau_w}\right)}-1}{\sqrt{\left(1+\frac{\alpha y}{\tau_w}\right)}+1}\right]+c$$

$$\text{and } B=\frac{3}{2}b_1\frac{\alpha}{\tau_w}\int\left(1+\frac{\alpha y}{\tau_w}\right)^{-\frac{1}{2}}dy$$

$$=\frac{3}{2}b_1\frac{\alpha}{\tau_w}\left[\left(1+\frac{\alpha y}{\tau_w}\right)^{\frac{1}{2}}\times\frac{1}{\frac{1}{2}}\times\frac{1}{\frac{\alpha}{\tau_w}}\right]$$

$$=3\,b_1\left(1+\frac{\alpha y}{\tau_w}\right)^{\frac{1}{2}}$$

The velocity profile with strong pressure gradient becomes:

$$\therefore \frac{\bar{U}}{u^*}=3\,b_1\left(1+\frac{\alpha y}{\tau_w}\right)^{\frac{1}{2}}+\frac{1}{k}\ln\left[\frac{\sqrt{\left(1+\frac{\alpha y}{\tau_w}\right)}-1}{\sqrt{\left(1+\frac{\alpha y}{\tau_w}\right)}+1}\right]+c \qquad \textbf{(3.28)}$$

As T.B.L. tend to separation $\frac{\alpha y}{\tau_w} \rightarrow \infty$

Equation (3.28) reduces to

$$\frac{\bar{U}}{u^*} = 3\, b_1 \sqrt{\left(1 + \frac{\alpha y}{\rho u^{*2}}\right)} + c \tag{3.29}$$

Equation (3.29) is the velocity profile of separating turbulent B.L.

CHAPTER 4

Transient Flow

Abstract: The study of hydraulic transients began with the investigation of the propagation of sound waves in air, the propagation in shallow water and flow of blood in arteries. However, none of these problems could solved rigorously until the development of elasticity and calculus and the solution of parties' differential equations.

In this chapter, a number of commonly used terms are defined, and a brief history of the development of the knowledge of hydraulic transients is presented.

The basic water hammer equations for the change in pressure caused by an instantaneous change in flow velocity are then derived. A description of the propagation and reflection of waves produced by closing value at downstream and of a single pipeline is presented. This is followed by a discussion of the classification and causes of hydraulic transients.

Unsteady flow through closed conduits is described by the dynamic and continuity equations. The derivation of these equations is presented and methods available for their solution are discussed.

Methods for controlling transients flow using various devices available to reduce or to eliminate the undesirable transients. Boundary conditions for these devices will be developed, which are required for the analysis of a system by the method of characteristics.

Keywords: Continuity equations, Controlling transients flow, Hydraulic transients, Propagation and reflection of waves, Water hammer.

4.1. DEFINITIONS

4.1.1. Transient State or Transient Flow

The intermediate stage flow, when the flow conditions are change from one steady – state condition to another steady state is called transient – state flow [10].

Jafar Mehdi Hassan, Riyadh S. Al-Turaihi, Salman Hussien Omran, Laith Jaafer Habeeb, Alamaslamani Ammar Fadhil Shnawa

4.1.2. Steady – Oscillatory or Periodic Flow

It is the flow conditions are varying with time and if they repeat after a fixed time interval the flow is called steady oscillatory flow and the time interval at which conditions are repeating is referred to as the period.

4.1.3. Column Separation

If the pressure in a closed conduit drops below the vapor pressure of the liquid, then cavitation formed in the liquid and the liquid column may separate.

4.1.4. Water Hammer

In the past, term water – hammer, oil hammer and steam hammer referred to the pressure caused by a flow change depending upon the fluid involved. Nowadays, the term hydraulic transient is used more frequently.

4.2. PRESSURE CHANGES CAUSED BY AN INSTANTANEOUS VELOCITY CHANGE

Let as consider the piping system of Fig. (**4.1**) in which a fluid is flowing with velocity V_o and pressure P_o. If the valve setting is changed instantaneously at a time $t = 0$, the velocity change to $V_o + \Delta V$ at the pressure $P_o + \Delta P$ the density ρ_o to $\rho_o + \Delta\rho$ and pressure wave of magnitude ΔP travels in the upstream direction. Let as designate the velocity of propagation of the pressure wave by (a), to simplify the derivation, let as assume that the pipe is rigid *i.e.* the diameter does not change due to pressure changes.

The unsteady flow situation is converted into steady conditions by superimposing on the C.V. wave velocity (a) in the downstream.

- The rate of change of momentum in the position of positive direction:

$$= \dot{m}\,\overrightarrow{\Delta V} = \rho Q\,\overrightarrow{\Delta V}$$

$$= \rho_o (V_o + a)\,A[(V_o + \Delta V + a) - (V_o + a)]$$

$$\Sigma F = \rho_o (V_o + a)\,A\,\Delta V \tag{4.1}$$

Neglecting friction, the resultant for $\sum F$ acting on the fluid in the C.V. in positive x – dir.

And

$$\sum F = P_o A - (P_o + \Delta P) A \tag{4.2}$$

i.e. from eqs. (4.1, 4.2)

$$\Delta P = -\rho_o (V_o + a)\, \Delta \tag{4.3}$$

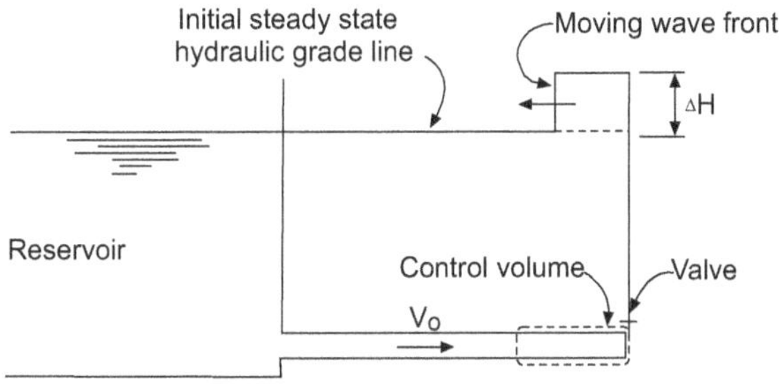

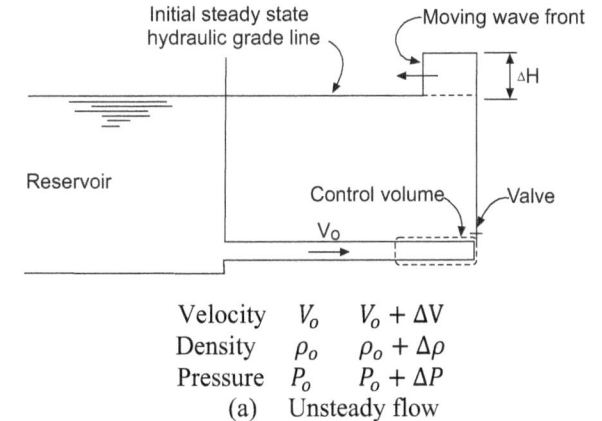

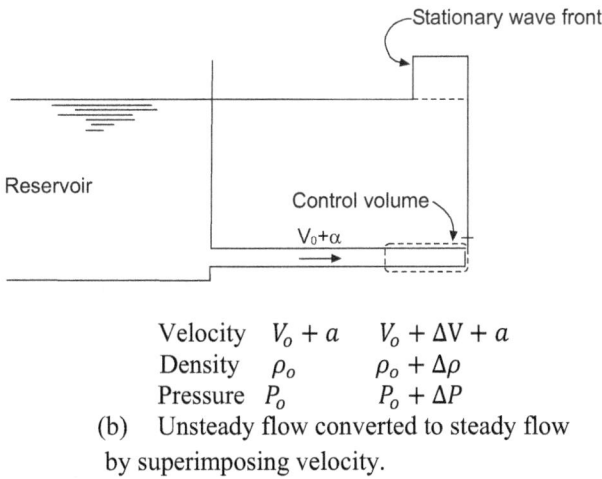

Velocity	V_o	$V_o + \Delta V$
Density	ρ_o	$\rho_o + \Delta\rho$
Pressure	P_o	$P_o + \Delta P$

(a) Unsteady flow

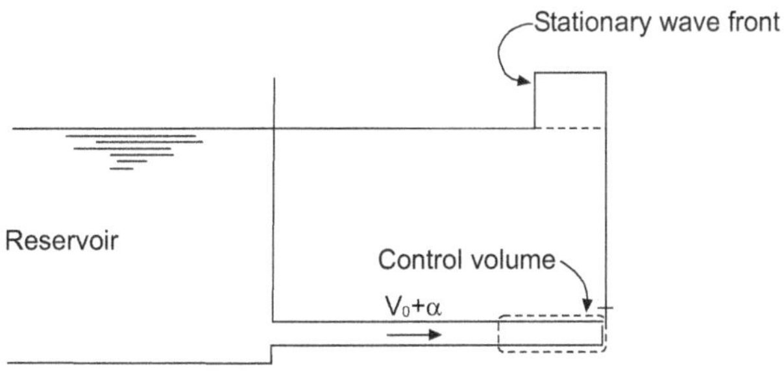

Velocity	$V_o + a$	$V_o + \Delta V + a$
Density	ρ_o	$\rho_o + \Delta\rho$
Pressure	P_o	$P_o + \Delta P$

(b) Unsteady flow converted to steady flow by superimposing velocity.

Fig. (4.1). Pressure rise in a pipeline due to instantaneous reduction of velocity ($\Delta P = \rho g \Delta H$) [10].

Since the wave velocity in metal or concrete pipes on in the rock tunnels (a) (approximately $1000\, m/s$) is much greater the $V_o (< 10\ m/s)$. Hence V_o is neglected [1].

Also

$$P = \rho g H$$

$$\Delta P = -\rho_{\circ}\, a \Delta V \quad \textbf{(4.4)}$$

And

$$\Delta H = -\frac{a}{g}\, \Delta V \quad \textbf{(4.5)}$$

The negative sign mean that the wave moving in the upstream direction and $\Delta H \propto \frac{1}{\Delta V}$

$$\text{also} \quad \Delta H = \frac{a}{g}\, \Delta V \quad \textbf{(4.6)}$$

The wave moving downstream

$$\Delta H \propto \Delta V.$$

- Rate of mass in flow $= \rho_o\, A(V_o + a)$
- Rate of mass out flow $= (\rho_o + \Delta\rho) A (V_o + \Delta V + a)$

The increase in the mass due density change is small and may be neglected.

$$\rho_o\, A(V_o + a) = (\rho_o + \Delta\rho) A (V_o + \Delta V + a)$$

Which upon simplification becomes:

$$\Delta V = -\frac{\Delta\rho}{\rho_o} (V_o + \Delta V + a)$$

Since $(V_o + \Delta V) \ll a$

$$\therefore \Delta V = -\frac{\Delta \rho}{\rho_o}\, a \tag{4.7}$$

The Bulk modulus of elasticity *K*, of fluid defined as

$$K = -\frac{\Delta P}{\Delta V / V_o} = \frac{\Delta P}{\Delta \rho / \rho_o} \tag{4.8}$$

From eqs. (4.7, 4.8)

$$a = -K\frac{\Delta V}{\Delta P}$$

Since $\Delta P = -\rho_o\, a\Delta V$

$$\therefore a = \sqrt{\frac{K}{\rho_o}} \tag{4.9}$$

This expression for rigid wall.

Example

Compute the velocity of pressure wave in 0.5 m dia. Pipe conveying oil from a reservoir to valve. Determine the pressure rise if a steady flow of 0.4 m^3/s is instantaneously measured at the downstream end by closing the valve.

Solution:

Assume the pipe is rigid, $\rho = 900\ KN/m^3$

$$K = 1.5\ GPa$$

$$A = \frac{\pi}{4}(0.5)^2 = 0.196\ m^2$$

$$V = \frac{Q}{A} = \frac{0.4}{0.196} = 2.04\,m/s$$

$$a = \sqrt{\frac{K}{\rho}} = \sqrt{\frac{1.5 \times 10^9}{900}} = 1291\ m/s$$

As the flow completely stopped, $\Delta V = 0 - 2.04\ m/s$

Therefore

$$\Delta H = -\frac{a}{g}\,\Delta V = -\frac{1291}{9.81}(-2.04)$$

$$= 268.5\,m$$

Since the sign of ΔH positive, it is a pressure rise.

4.3. WAVE PROPAGATION AND REFLECTION IN A SINGLE PIPELINE

Fig. (**4.2**) shows a piping system which flow conditions are steady and at time too, the valve is instantaneously closed.

Intial pressure head H_o, let the distance x and the velocity V be events following the valve closure can be divided into four part as follows:

1. $0 < t \leq \frac{L}{a}$ Fig. **(4.2a, b)**

As soon as the valve is closed the flow velocity at the valve is reduced to zero, which cause the pressure rise of

$$\Delta H = \frac{a}{g} V_o.$$

Because of this Pressure rise, the pipe expands the fluid compressed, thus increasing density, and positive pressure wave propagate toward the reservoir. Behind this wave, the flow velocity is zero, and all the K.E. converted into elastic energy. If a is the wave velocity and L is the length of pipeline, then the time $t = \frac{L}{a}$, along the entire length of the pipeline, the pipe expanded, the flow velocity is zero, and the pressure head is $H_o + \Delta H_o$. The flow will fluctuated until becomes stead state.

Fig. (**4.2**) illustrate the sequence of events along the pipeline as follows:

2. $\frac{L}{a} < t \leq \frac{2L}{a}$ (c and d)

3. $\frac{2L}{a} < t \leq \frac{3L}{a}$ (e, f)

4. $\frac{3L}{a} < t \leq \frac{4L}{a}$ (g, h)

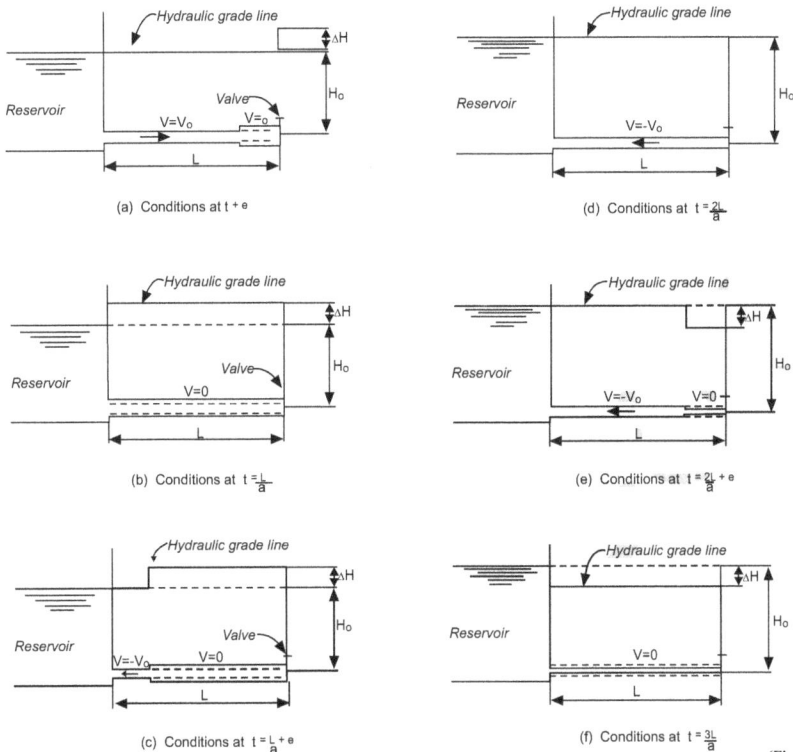

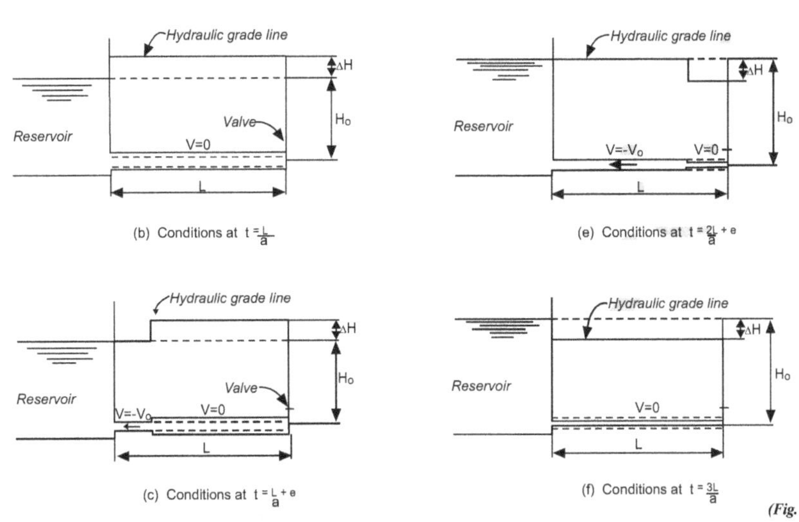

(a) Conditions at t + e

(d) Conditions at $t = \frac{2L}{a}$

(b) Conditions at $t = \frac{L}{a}$

(e) Conditions at $t = \frac{2L}{a} + e$

(c) Conditions at $t = \frac{L}{a} + e$

(f) Conditions at $t = \frac{3L}{a}$

(Fig. 4.2) contd.....

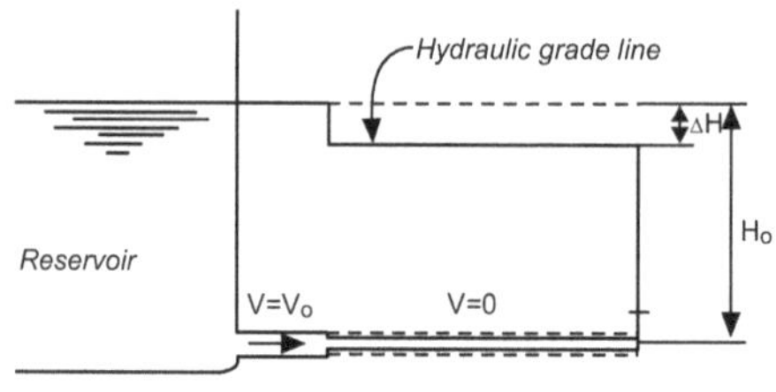

(g) Conditions at $t = \frac{3L}{a} + e$

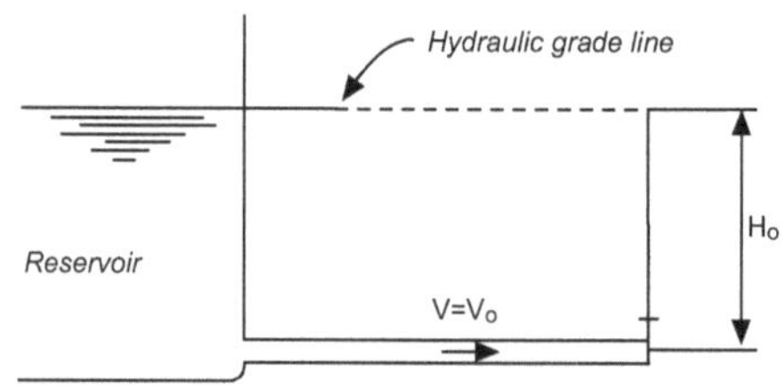

(h) Conditions at $t = \frac{4L}{a}$

Fig. (4.2). Propagation of pressure waves caused by instantaneous closure of valve [10].

Fig. (**4.3**) shows the pressure variation at the valve end with time. As we assumed the system is frictionless, this process conditions and the conditions are repeated at an interval of 4L/a. this interval after which conditions are repeated is termed the theoretical period of the pipeline.

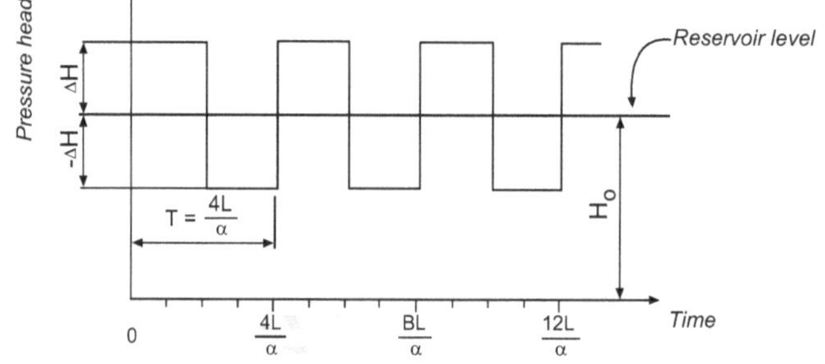

Fig. (4.3). Pressure variation at valve; friction losses neglected [10].

In real physical systems, however, pressure waves are dissipated due to friction losses as the waves propagate in the pipeline, and the fluid becomes stationary after a short time.

If the friction losses are taken into consideration, then the pressure variation at the valve with time will be as shown in Fig. (**4.4**).

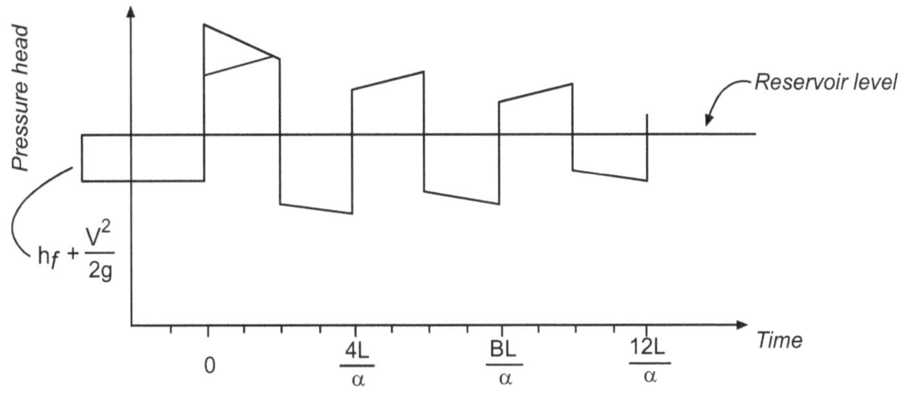

Fig. (4.4). Pressure Variation at valve; friction losses considered [10].

4.4. CLASSIFICATION OF HYDRAULIC TRANSIENTS

Depending upon the conduits which the transient conditions are occurring transients may be classification three categories [11].

1. Transients in closed conduit.
2. Transients in open channel.
3. Combined free – surface pressurized transient flow.

4.5. CAUSES OF TRANSIENTS

1. Opening, closing or "chattering" of valves in a pipeline.
2. Starting or stopping the pumps in a pumping system.
3. Starting up a hydraulic turbine, accepting or rejecting load.
4. Vibrations of the vanes of a runner or an impeller, or of the blades of a fan.
5. Sudden changes in the inflow or outflow a canal by opening or closing the control gate.
6. Failure or collapse of a dam.

7. Sudden increases in the inflow to a river on a sewer due to flush storm runoff.

4.6. EQUATIONS OF UNSTEADY FLOW THROUGH CLOSED CONDUITS

Unsteady flow through closed conduits is described by the dynamic and continuity equations. In this chapter, the derivation of these equations is presented, and methods available for their solution are discussed [10].

4.6.1. Assumptions

The following assumptions are made in the derivation of the equations:

1. Flow in the conduit is one – dimensional, and the velocity distribution is uniform over the cross section of the conduit.

2. The conduit walls and the fluid are linearly elastic, *i.e.*, stress is proportional to strain. This is true for most conduits such as metal, concrete and wooden pipes, and lined or unlined rock tunnels.

3. Formulas for computing the steady – state friction losses in conduits are valid during the transient state. The validity of this assumption has not as yet been verified. For computing frequency – dependent friction, it has been developed a procedure for laminar flows, and proposed an empirical procedure for turbulent flows. However, these procedures are too complex and cumbersome for general use, and we will not discuss them further.

4.6.2. Dynamic Equation

We will use the following notation: distance, x, discharge, Q, and flow velocity, V_o are considered positive in the downstream direction (see Fig. **4.5**), and H is the piezometric head at the centerline of the conduit above the specified datum. Let us consider a horizontal element of fluid having cross – sectional area A and length δx, within a conduit as shown in Fig. (**4.5**). If the piezometric head and the velocity at distance x are H and V, then their corresponding values at $x + \delta x$ are $H + (\partial H/\partial x)\delta x$ and $V + (\partial V/\partial x)\delta x$, respectively. In the x – direction, three [1, 10].

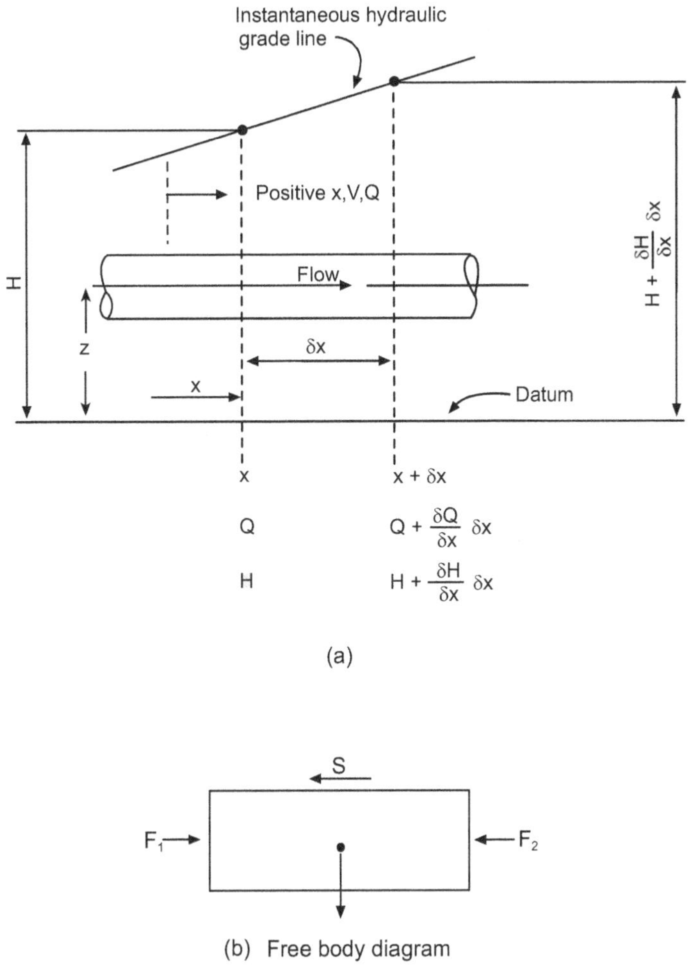

Fig. (4.5). Notation for dynamic equation [10].

Forces, F_1, F_2, and S, are acting on the element. F_1 and F_2 are forces due to pressure while S is the shear force due to friction. If γ = specific weight of the fluid, A = cross – sectional area of conduit, and z = height of conduit above datum, then

$$F_1 = \gamma A(H - z) \quad \textbf{(4.10)}$$

$$F_2 = \left(H - z + \frac{\partial H}{\partial x}\delta x\right) A \quad \textbf{(4.11)}$$

If the Darcy – Weisbach formula is used for computing the friction losses, then the shear force

$$S = \frac{y}{g}\frac{fV^2}{g}\pi D\delta x \quad \textbf{(4.12)}$$

In which g = acceleration due to gravity, f = friction, and D = diameter of the conduit. The resultant force, F, acting on the element is given by the equation

$$F = F_1 - F_2 - S \quad \textbf{(4.13)}$$

Substitution of the expressions for F_1, F_2, and S_1 from Eqs. 4.10 through 4.12 into Eq. 4.13 yields

$$F = -\gamma A\frac{\partial H}{\partial x}\delta x - \frac{\gamma}{g}\frac{fV^2}{g}\pi D\delta x \quad \textbf{(4.14)}$$

According to Newton's second law of motion,

$$Force = Mass \times Acceleration. \quad \textbf{(4.15)}$$

For the fluid element under consideration,

$$\left.\begin{aligned} Mass\ of\ the\ element &= \frac{\gamma}{g}A\delta x \\ Acceleration\ of\ the\ element &= \frac{dV}{dt} \end{aligned}\right\} \quad \textbf{(4.16)}$$

Substitution of Eqs. 4.14 and 4.16 into Eq. 4.15 and division by $\gamma A\delta x$ yield

$$\frac{dV}{dt} = -g\frac{\partial H}{\partial x} - \frac{fV^2}{2D} \quad \textbf{(4.17)}$$

We know from elementary calculus that the total derivative

$$\frac{dV}{dt} = \frac{\partial V}{\partial t} + \frac{\partial V}{\partial x}\frac{\partial x}{\partial t} \quad \textbf{(4.18a)}$$

Or

$$\frac{dV}{dt} = \frac{\partial V}{\partial t} + V\frac{\partial V}{\partial x} \quad \textbf{(4.18b)}$$

Substituting Eq. 4.18b into Eq. 4.17 and rearranging,

$$\frac{dV}{dt} + V\frac{\partial V}{\partial x} + g\frac{\partial H}{\partial x} + \frac{fV^2}{2D} = 0 \quad \textbf{(4.19)}$$

In most of the transient problem,[7] the term $V\,\partial V/\partial x$ is significantly smaller than the term $\partial V/\partial t$. Therefore, the former may be neglected. To account for the reverse flow, the expression V^2 in eq. 4.19 may be written as $V|V|$, in which $|V|$is the absolute value of V. By writing Eq. 4.19 in term of discharge, *Q*, and rearranging,

we obtain

$$\frac{dQ}{dt} + gA\frac{\partial H}{\partial x} + \frac{fV^2}{2DA}Q|Q| = 0 \quad \textbf{(4.20)}$$

In Eqs. 4.12, 4.14, 4.17, 4.19, and 4.20, the Darcy – Weisbach has been used for calculating the friction losses. If a general exponential formula had been used for these losses, then the last term of Eq. 4.20 could be written as $kQ|Q|^m D^b$, with the values of k, m and b depending upon the formula employed. For example, for the Hazen – Williams formula, m = 1.85 and b = 2.87 while, as derived above for the Darcy – Weisbach formula m = 1 and b = 3 If correct values of m and b are used, the results are independent of the formula employed, *i.e.*, the Darcy – Weisbach and the Hazen – Williams formulas would give comparable results.

4.6.3. Continuity Equation

Let us consider the control volume shown in Fig. (**4.6**). The volume of fluid inflow, $\forall_{in}$, and outflow, $\forall_{out}$, during time interval δt are [1,10].

$$\forall_{in} = V\pi r^2 \delta t \quad \textbf{(4.21)}$$

$$\forall_{out} = \left(V + \frac{\partial V}{\partial x}\delta x\right)\pi r^2 \delta t \quad \textbf{(4.22)}$$

In which r = radius of the conduit. The increase in the fluid volume, $\delta\forall_{in}$, during time δt

$$\delta\,\forall_{in} = \forall_{in} - \forall_{out} = -\frac{\partial V}{\partial x}\delta x\;\delta t\;\pi r^2 \quad \textbf{(4.23)}$$

The pressure change, δp, during time interval δt is $(\partial p/\partial t)\delta t$. This pressure change causes the conduit walls to expand or contract radially and cause the length of the fluid element to decrease or increase due to fluid compressibility (see Fig. **4.6**).

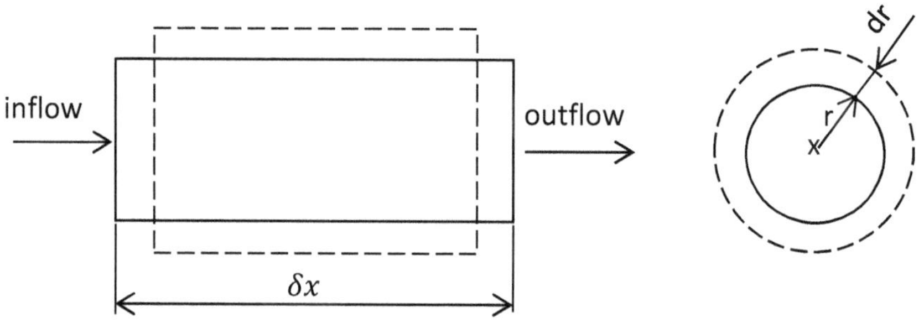

Fig. (4.6). Notation for continuity equation [10].

Let us first consider the volume change, $\delta \forall_r$, due to the radial expansion or contraction* of the conduit. The radial or hoop stress, σ, in a conduit due to the pressure p is given by the equation

$$\sigma = \frac{pr}{e} \tag{4.24}$$

In which e = the conduit wall thickness. Hence, the change in hoop stress, $\delta\,\sigma$, caused by δp may be written as:

$$\delta\sigma = \delta p\frac{r}{e} = \frac{\partial p}{\partial t}\delta t\frac{r}{e} \tag{4.25}$$

Since the radius r has increased to $r + \delta r$, the change in strain

$$\delta\epsilon = \frac{\delta r}{r} \tag{4.26}$$

If the conduit walls are assumed linearly elastic, then

$$E = \frac{\delta\sigma}{\delta\epsilon} \tag{4.27}$$

In which E = Young's modulus of elasticity. Substitution of expressions for $\delta\sigma$ and $\delta\epsilon$ from Eqs. 4.25 and 4.26 into Eq. 4.27 yields.

$$E = \frac{(\partial p/\partial t)\delta t(r/e)}{\delta r/r} \quad \textbf{(4.28)}$$

Or

$$\delta r = \frac{\partial p}{\partial t}\frac{r^2}{eE}\delta t \quad \textbf{(4.29)}$$

The change in the volume of the element due to the radial expansion or contraction of the conduit is

$$\delta \forall_r = 2\pi r\ \delta x\ \delta r \quad \textbf{(4.30)}$$

Substituting for δr from Eq. 4.29 yields

$$\delta \forall_r = 2\pi\frac{\partial p}{\partial t}\frac{r^3}{eE}\ \delta t\delta x \quad \textbf{(4.31)}$$

Let us now derive an expression for the change in volume, $\delta \forall_c$, due to compressibility of the fluid. The initial volume of the fluid element.

$$\forall = \pi r^2\ \delta x \quad \textbf{(4.32)}$$

The bulk modulus of elasticity of a fluid, *K*, is defined[1] as

$$K = \frac{-\delta p}{\delta\forall_c/\forall} \quad \textbf{(4.33)}$$

By substituting for $\forall$ from Eq. 4.32 and noting that

$\delta p = (\partial p/\partial t)\delta t$, Eq. 4.33 becomes

$$\delta \forall_c = \frac{-\partial p}{\partial t}\frac{\delta t}{K}\pi r^2 \delta x \quad \textbf{(4.34)}$$

If we assume that the fluid density remains constant, then it follows from the law of conservation of mass that

$$\delta \forall_{in} + \delta \forall_c = \delta \forall_r. \quad \textbf{(4.35)}$$

Substitution of expression for $\delta \forall_{in}, \delta \forall_r$, and $\delta \forall_c$ from Eqs. 4.23, 4.31, and 4.34 into the above equation and division by $\pi r^2 \delta x\, \delta t$ yield

$$-\frac{\partial V}{\partial x} - \frac{1}{K}\frac{\partial p}{\partial t} = \frac{2r}{eE}\frac{\partial p}{\partial t} \tag{4.36}$$

Or

$$\frac{\partial V}{\partial x} + \frac{\partial p}{\partial t}\left(\frac{2r}{eE} + \frac{1}{K}\right) = 0 \tag{4.37}$$

Let us define

$$a^2 = \frac{K}{\rho[1+(KD/eE)]} \tag{4.38}$$

In which ρ = mass density of the fluid. Noting that $p = \rho g H$, rearranging the terms, and substituting $Q = VA$, Eq.4.37 becomes

$$\frac{a^2}{gA}\frac{\partial Q}{\partial x} + \frac{\partial H}{\partial t} = 0 \tag{4.39}$$

It will be shown in the next section that a is the velocity of waterhammer waves. Expressions for a for various conduit and support conditions are presented in Section 4.7.

4.7. Velocity of Waterhammer Waves

An expression for the velocity of waterhammer waves in a rigid conduit was derived in Section 4.2. however, in addition to the bulk modulus of elasticity, K, of the fluid, the velocity of waterhammer waves depends upon the elastic properties of the conduit, as well as on the external constrains, Elastic properties include the conduit size, wall thickness, and wall material; the external constraints include the type of supports and the freedom of conduit movement in the longitudinal direction. The bulk modulus of elasticity of a fluid depends upon its temperature, pressure, and the quantity of undissolved gases. Since the wave velocity changes by about 1 percent per 5 ℃. The fluid compressibility is increased by the presence of free gases, and it has been found that 1 part of air in 10,000 parts of water by volume reduces the wave velocity by about 50 percent. *

Solids in liquid have similar but less drastic influence, unless they are compressible. Laboratory and prototype tests have shown that the dissolved gases tend to come out of solution when the pressure is reduced, even when it remains above the vapor pressure. This cause a significant reduction in the wave velocity. Therefore, the wave velocity for a positive wave may be higher than that of a negative wave. Further prototype tests are needed to quantify the reduction in the wave velocity due to reduction of pressures.

From the literature, the following general expression for the wave velocity:

$$a = \sqrt{\frac{K}{\rho[1+(K/E)\psi]}} \tag{4.40}$$

In which ψ is a nondimensionalized parameter that depends upon the elastic properties of the conduit; E = Young's modulus of elasticity of the conduit wall; and K and ρ are the bulk modulus of elasticity and density of the fluid, respectively. The moduli of elasticity of commonly used materials for conduit walls and the bulk moduli elasticity and mass densities of various liquids are listed in Tables **4.1** and **4.2**.

Expression for ψ for various conditions are follows:

1. Rigid Conduits:

$$\psi = 0 \tag{4.41}$$

2. Thick – Walled Elastic Conduits:

a) Conduit anchored against longitudinal movement throughout its length

$$\psi = 2(1+v)\frac{R_0^2+R_i^2}{R_0^2-R_i^2} - \frac{2vR_i^2}{R_0^2-R_i^2} \tag{4.42}$$

in which v = the Poisson's ratio and R_0 and R_i = the external and internal radii of the conduit.

Table 4.1: Young's modulus of elasticity and Poisson's ratio for various pipe materials [10]. *

Material	Modulus of Elasticity, E** (GPa)	Poisson's Ratio
Aluminum alloys	86 – 73	0.33
Asbestos cement, transit	24	
Brass	78 – 110	0.36
Cast iron	80 – 170	0.25
Concrete	14 – 30	0.1 – 0.15
Copper	107 – 131	0.34
Glass	46 – 73	0.24
Lead	4.8 – 17	0.44
Mild steel	200 – 212	0.27
Plastics		
ABS	1.7	0.33
Nylon	1.4 – 2.75	
Perspex	6.0	0.33
Polyethylene	0.8	0.46
Polystyrene	5.0	0.4
PVC rigid	2.4 – 2.75	
Rocks		
Granite	50	0.28
Limestone	55	0.21
Quartzite	24.0 – 44.8	
Sandstone	2.75 – 4.8	0.28
Schist	6.5 – 18.6	

**To convert *E* into Ib/in.^{2}, multiply the values given in this column by 145.038 × 10^3.

b) Conduit anchored against longitudinal movement at the upper end

$$\psi = 2\left[\frac{R_0^2+1.5\,R_i^2}{R_0^2-R_i^2}+\frac{v\left(R_0^2-3\,R_i^2\right)}{R_0^2-R_i^2}\right] \quad \textbf{(4.43)}$$

c) Conduit with frequent expansion joints

$$\psi = 2\left(\frac{R_0^2 + R_i^2}{R_0^2 - R_i^2} + v\right) \quad \textbf{(4.44)}$$

3. Thin – Walled Elastic Conduits:

a) Conduit anchored against longitudinal movement throughout its length

$$\psi = \frac{D}{e}(1 - v^2) \quad \textbf{(4.45)}$$

in which D = conduit diameter and e = wall thickness.

Table 4.2: Bulk modulus of elasticity and density of common liquids at atmospheric pressure [11].*

Liquid	Temperature (°C)	Density, ρ^+ (kg/m³)	Bulk Modulus of Elasticity, $K^‡$ (GPa)
Benzene	15	880	1.05
Ethyl alcohol	0	790	1.32
Glycerin	15	1,260	4.43
Kerosine	20	804	1.32
Mercury	20	16,570	26.2
Oil	15	900	1.5
Water, fresh	20	999	2.19
Water, sea	15	1,025	2.27

*To determine the specific weight of the liquid, in Ib_f/ft^3, multiply the values given in this column by 62.427 × 10^{-3}.

‡ To convert K into Ib/in.², multiply the values given in this column by 145.038 × 10^3.

b) Conduit anchored against longitudinal movement at the upper end length

$$\psi = \frac{D}{e}(1.25 - v) \quad \textbf{(4.46)}$$

c) Conduit with frequent expansion joints

$$\psi = \frac{D}{e} \quad \textbf{(4.47)}$$

4. Tunnels Through Solid Rock:

It has been derived long expressions for ψ at the literature for lined and unlined rock tunnels. Usually, the rock characteristics cannot be precisely estimated because of nonhomogeneous rock conditions and because presence of fissures. Therefore, in our opinion, using an expression for practical applications unwarranted. Instead, the following expressions based on

a) Unlined tunnel

$$\left.\begin{matrix}\psi=1\\E=G\end{matrix}\right\} \quad \textbf{(4.48)}$$

in which G = modulus of rigidity of the rock.

b) Steel – lined tunnel

$$\psi=\frac{DE}{GD+Ee} \quad \textbf{(4.49)}$$

In which e = thickness of the steel liner and E = modulus of elasticity of steel.

5. Reinforced Concrete Pipes:

The reinforced concrete pipe is replaced by an equivalent steel pipe having equivalent thickness

$$e_e = E_r e_c + \frac{A_s}{I_s} \quad \textbf{(4.50)}$$

In which e_c = thickness of the concrete pipe; A_s and I_s are the cross – sectional area and the spacing of steel bars, respectively; and E_r = ratio of the modulus of elasticity of concrete to that of steel. Usually the value of E_r varies from 0.06 to 0.1. However, to allow for any cracks in the concrete pipe, a value of 0.05 is suggested. Having computed e_e, the wave velocity may be determined from Eq. 4.40 using the modulus elasticity of steel.

6. Wood – Stave Pipes:

The thickness of a uniform steel pipe equivalent to the wood – stave pipe is determined from Eq. 4.50 using $E_r = \frac{1}{60}, e_c =$ thickness of wood staves, and A_s and I_s are the cross – sectional area and the spacing of the steel bands, respectively.

The wave velocity is then computed from Eq. 4.40.

7. Polyvinyl Chloride (PVC) and Reinforced Plastic Pipe:

Investigations reported show that Eq. 4.40 can be used for computing wave velocity in the polyvinyl chloride (PVC) and reinforced plastic pipes, provided a proper value of the modulus of elasticity for the wall material is used.

8. Noncircular Conduits:

The following expression for ψ is obtained from the equation for the wave velocity in the thin – walled rectangular conduits by using the steady – state bending theory and by allowing the corners of the conduit to rotate:

$$\psi = \frac{\beta b^4}{15\, e^3 d} \tag{4.51}$$

In which $\beta = 0.5\,(6 - 5\alpha) + 0.5\,(d/b)^3\,[6 - 5(b/d)^2]$, $\alpha = \frac{[1+(d/b)^3]}{[1+(d/b)]}$, $b =$ width of the conduit (longer side), and $d =$ depth of the conduit (shorter side).

4.8. METHODS FOR CONTROLLING TRANSIENTS

4.8.1. General

A piping system can be designed with a liberal factor of safety to withstand the maximum and minimum pressure caused by any possible operating condition expected to occur during the life of the system. Such a design in most cases will, however, be very uneconomical. Therefore, various devices and/or control procedures are used to reduce or eliminate undesirable transients, *e.g.*, excessive pressure rise or drop, column separation, pump or turbine over speed. Such devices are usually costly, and there is no single device that is suitable for all system or for all operating conditions. Therefore, while designing a piping system, a number of alternatives should be considered. The alternative that gives an acceptable system response and an overall economic system should be selected. An acceptable system response may be defined by specifying limits on the maximum and minimum pressure, maximum turbine speed following full – load rejection, or maximum reverse pump speed following power failure [10, 11].

In section 4.2, we derived the following equation for pressure change, ΔH, as a result of an instantaneous change in the flow velocity, ΔV,

$$\Delta H = -\frac{a}{g}\Delta V \tag{4.52}$$

In which a = waterhammer wave velocity and g = acceleration due to gravity. This equation indicates that the main function of a device used for the reduction of the magnitude of pressure rise or pressure drop would be to reduce ΔV and/or a. In addition, the flow velocity, V, may be varied in such a manner that the pressure is kept within the prescribed limits. Such *a* controlled variation of the flow conditions, which results in *a* required system response, is referred to as optimal control of transient flows.

4.8.2. Available Device and Methods for Controlling Transients

The following devices are commonly used to reduce or to eliminate the undesirable transients, such as excessive pressure, column separation, and pump or turbine overspeed following a power failure or a load rejection: (1) surge tanks, (2) air chambers, and (3) valves.

In addition, the severity of undesirable transient may be reduced by changing the pipeline profile, by increasing the diameter of the pipeline, or by reducing the waterhammer wave velocity.

4.8.3. Surge Tanks

A surge tank is an open standpipe or a shaft connected to the conduits of a hydroelectric power plant or to the pipeline of a piping system. This is also referred to as a surge shaft or surge chamber.

The main functions of a surge tank are:

1. It reduces the amplitude of pressure fluctuations by reflecting the incoming pressure waves. For example, the waterhammer waves produced in a penstock by load changes on a turbine (Fig. **4.7**) are reflected back at the surge tank. Thus, the conduit length to be used in the waterhammer analysis is between the turbine and the surge tank rather than between the turbine and the upstream reservoir. Due to this reduction in the conduit length, the pressure rise or drop is less than if the surge tank were not provided. In addition, if a surge tank were not present at the junction of the penstock and the tunnel, then the tunnel would have to be designed to withstand the waterhammer pressure.
2. A surge tank improves the regulating characteristics of a hydraulic turbine. Because of the surge tank, the length of the power conduit to be used for determining the water – starting time is up to the surge tank rather than up to the upstream reservoir. The water – starting time of a hydropower scheme is therefore reduced, thus improving the regulating characteristics of the power plant.
3. A surge tank acts as a storage for excess water during load reduction in a hydropower plant and during start – up of the pumps in a pumping system. Similarly, it provides water during load acceptance in a hydropower plant and during power failure in a pumping system. Therefore, the water is accelerated or decelerated in the pipeline slowly, and the amplitude of the pressure fluctuations in the system is reduced.

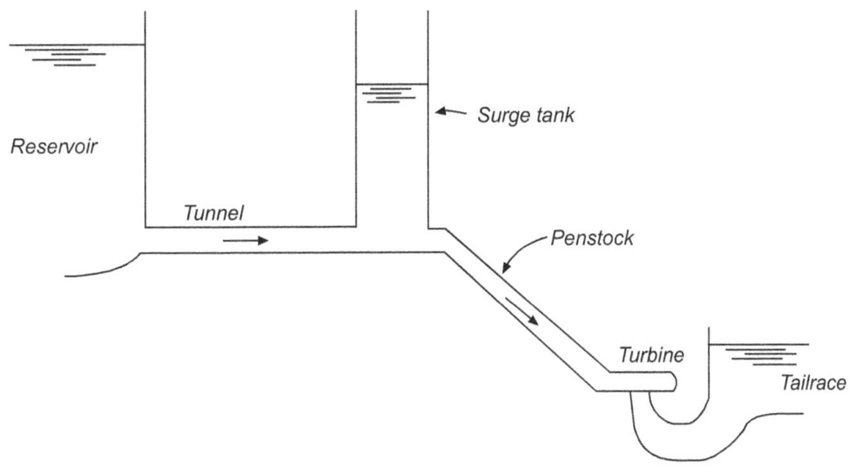

Fig. (4.7). Schematic diagram of a hydroelectric power plant [10].

4.8.4. Types of Surge Tanks

Depending upon its configuration, a surge tank may be classified as simple, orifice, differential, one – way, or closed. A brief description of each follows:

A simple surge tank is just a shaft or standpipe connected to the pipeline. If the entrance to the surge tank is restricted by means of an orifice, it is called an orifice tank. An orifice tank having a riser is termed differential. In a one – way surge tank, the liquid flows from the tank into the pipeline only when the pressure in the pipeline drops below the liquid level in the surge tank. Following the transient – state conditions, the tank is filled from the pipeline. If the top of the tank is closed or if there is a valve or orifice in the vent stack connecting tank to the outer atmosphere, the tank is called a closed surge tank. Depending upon the requirement that the tank must fulfill, a simple tank may have upper or lower galleries. Fig. (**4.8**) shows a number of typical surge tank.

If necessary, a combination of different types of surge tank may be provided in an installation.

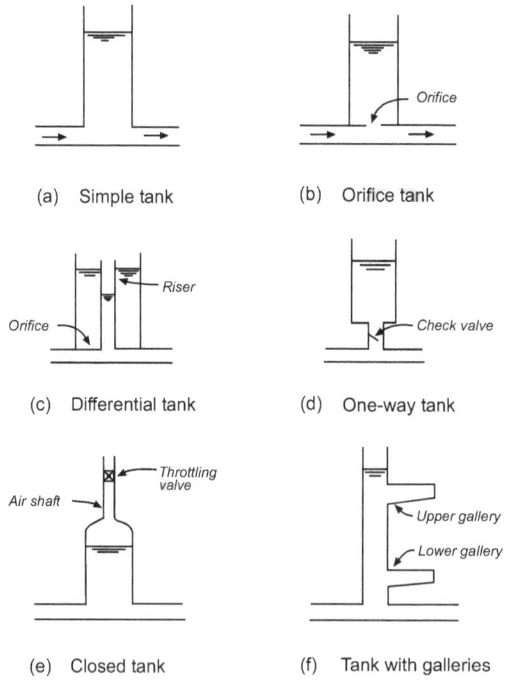

Fig. (4.8). Types of surge tanks [11].

4.8.5. Air Chambers

An air chamber(Fig. **4.9**) is a vessel having compressed air at its top and having liquid in its lower part. To restrict the inflow into or outflow from the chamber, an orifice is usually provided between the chamber and the pipeline. An orifice, which is shaped such that it produces more head loss for inflow into the chamber than for a corresponding outflow from the chamber, is referred to as differential orifice (Fig. **4.9**). To prevent very low minimum pressure in the pipeline and hence column separation, the outflow from the chamber should be as free as possible, while the inflow may be restricted to reduce the size of the chamber. A ratio of 2.5:1 between the orifice head losses for the same inflow and outflow is commonly used. As the air volume may be reduced due to leakage or due to solution in the liquid, an air compressor is used to keep the volume of the air within the prescribed limits.

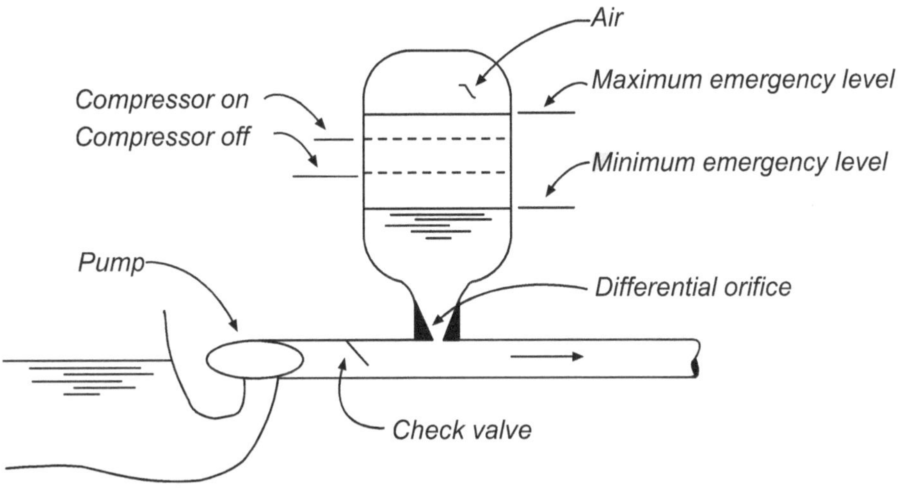

Fig. (4.9). Air chamber [11].

It is a common practice to provide a check valve between the pump and the air chamber (see Fig. **4.9**). Upon power failure, the pressure in the pipeline drops, and the liquid is supplied from the chamber into the pipeline. When the flow in the pipeline reverses, the check valve closes instantaneously, and the liquid flows into the chamber. Because of the inflow or outflow from the chamber, the air in the chamber contracts or expands, and the magnitude of the pressure rise and drop are reduced due to gradual variation of the flow velocity in the pipeline.

An air chamber has the following advantages over a surge tank:

1. The volume of an air chamber required for keeping the maximum and minimum pressure within the prescribed limits is smaller than that of an equivalent surge tank.
2. An air chamber can be installed with its axis parallel to the ground slope. This reduces the foundation costs and provides better resistance to both wind and earthquake loads.
3. An air chamber can be provided near the pump, which may not be practical in the case of a surge tank because of excessive height. This reduces the pressure rise and the pressure drop in the pipeline.
4. To prevent freezing in cold climates, it is cheaper to heat the liquid in an air chamber than in a surge tank because of smaller size and because of proximity to the pumphouse.

4.8.6. Valves

Depending upon the type, a valve is used to control the transients by either of the following operations:

1. The valve opens or closes to reduce the rate of net change in the flow velocity in the pipeline.
2. It allows rapid outflow of the liquid from the pipeline if the pressure exceeds a set limit. This outflow causes a pressure drop, thus reducing the maximum pressure.
3. The valve opens to admit air into the pipeline, thus preventing the pressure from dropping to the liquid vapor pressure.

A number of valve commonly used to control transients are:

1. Safety valves.
2. Pressure – relief valves.
3. Pressure – regulating valves.
4. Air – inlet valves.
5. Check valves.

A safety valve or an overpressure pop – off valve (Fig. **4.10a**) is a spring or weight – loaded valve, which opens as soon as the pressure inside the pipeline exceeds the pressure head set on the valve. The valve closes abruptly when the pressure drops below the limit set on the valve (Fig. **4.11a**), a safety valve is either fully open or fully closed.

The operation of a pressure – relief valve or surge suppressor (Fig. **4.10b**) is similar to that of a safety except that its opening is proportional to the amount by which the pressure in the pipeline just upstream of the valve exceeds the prescribed limits. The valve closes when the pipeline pressure drops and is fully closed when the pressure is below the limit set on the valve. There is usually some hysteresis in the opening and the closing of the valve, as shown in Fig. (**4.11b**).

For a pumping system having more than one pump discharging into a common header, a battery of smaller – size relief valves or surge suppressors may by used [13] instead of one large surge suppressor. A suppressor may be mounted on the main discharge line. In the latter arrangement, the overpressure setting of each valve should be set such that the valves open in sequence one after the other rather than simultaneously.

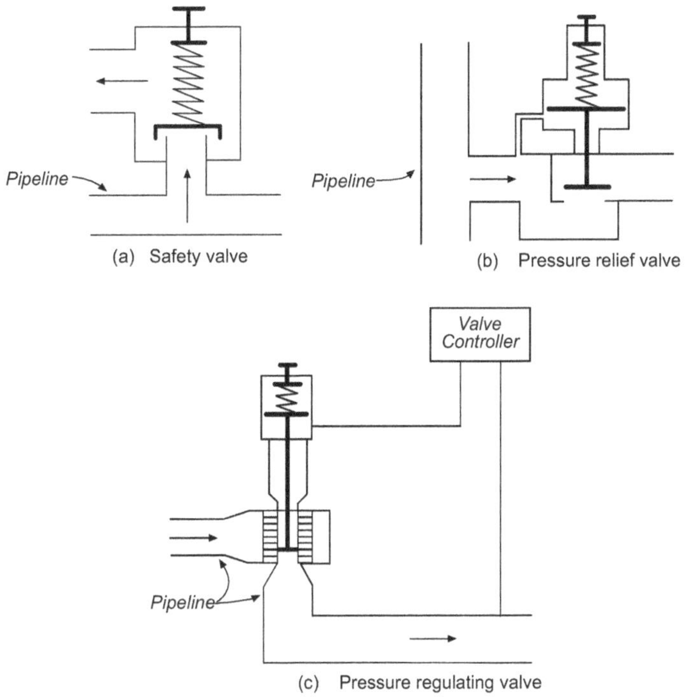

Fig. (4.10). Schematic diagrams of safety, relief, and pressure – regulating valves [10].

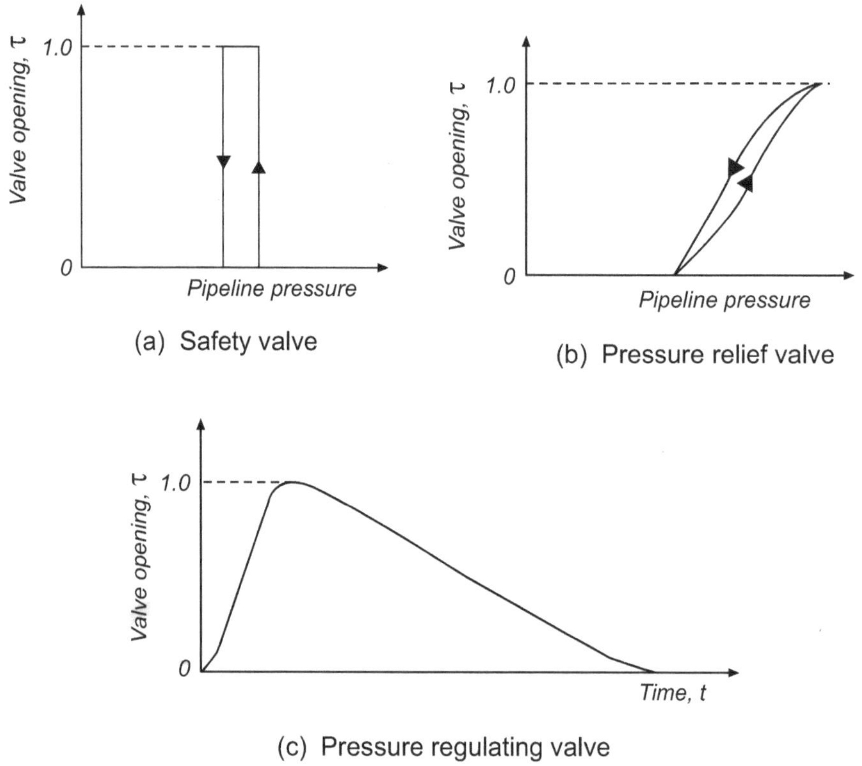

Fig. (4.11). Discharge characteristics of safety, relief, and pressure – regulating valves [10].

A pressure – regulating valve (PRV) is a pilot – controlled throttling valve, which is opened or closed by a servomotor, and the opening and closing times of this valve can be individually set. It is installed just downstream of a pump in a pumping system and upstream of a turbine in a hydropower scheme. Following power failure to the pump – motor, this valve rapidly opens and then gradually closes (Fig. **4.11c**) to reduce the pressure rise. The operation of this valve in a hydropower scheme is as follows:

If the power plant is isolated from the grid system, the PRV is kept partly open to provide for the maximum anticipated rapid load increase. When accepting rapid load changes, the PRV is closed at a slow rate. In such an operation, same water is wasted. However, as isolated operation is an emergency condition, the amount of water wasted is insignificant.

Upon full – load rejection, either in the normal or isolated operation of the turbine, all the turbine flow is switched from the turbine to the PRV, which is then closed slowly.

Fig. (**4.12**) illustrates the synchronous operation of a PRV and a turbine. Fig. (**4.12a** and **b**) are for a turbine isolated from the grid system. Fig. (**4.12c**) is for a turbine connected to or isolated from the grid system. Ideally, the net change in the penstock flow may be reduced to zero by matching the discharge characteristics of the PRV with that of the turbine. However, this is usually not possible because of the nonlinear flow characteristics of the turbine and valves and because of the dead or delay time between the opening (closing) of the pressure regulator and the closing (opening) of the wicket gates. This dead time should be as small as possible to minimize pressure rise or drop in the penstock.

Air – inlet valves are installed to admit air into the pipeline whenever the pressure inside the pipeline drops below the atmospheric pressure. Therefore, the pressure differential between the outside atmospheric pressure and the pressure inside the pipeline is reduced, thus preventing the collapse of the pipeline. Air inlet valves are also used to reduce the generation of high pressure when the liquid columns rejoin following column separation by providing an air cushion in the pipe.

Once air has been admitted into the pipeline, extreme care must be exercised while refilling the line. The air pockets should be eliminated gradually from the line because the entrapped air can result in very high pressure [14 – 17].

Check valves are used to prevent reverse flow through a pump and to prevent inflow into a one – way surge tank from the pipeline. They are installed immediately downstream of a pump or at the bottom of a one – surge tank. A check valve in its simplest from is a flap valve, although sometimes dashpots and springs are provided to prevent slamming of the valve.

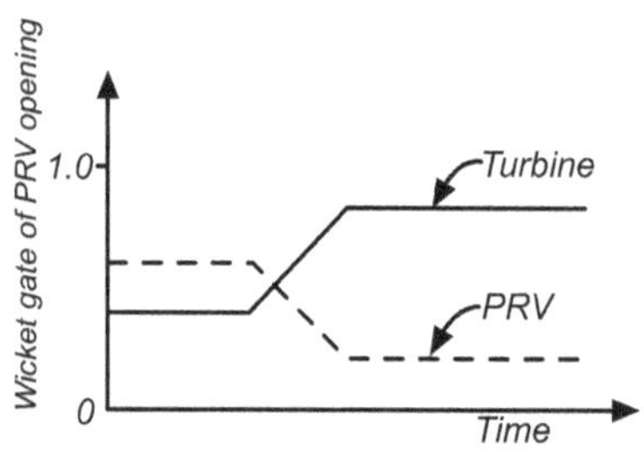

(a) Partial load acceptance in isolated operation

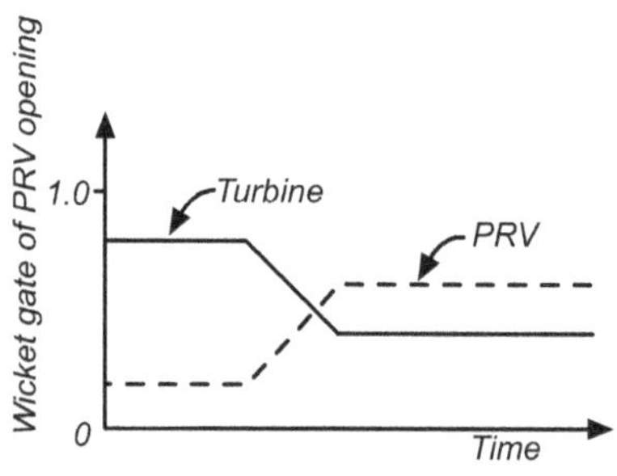

(b) Partial load rejection in isolated operation

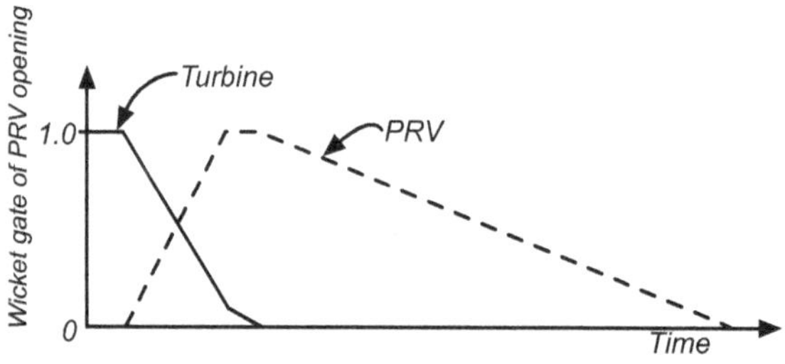

(c) Total load rejection in normal or isolated operation

Fig. (4.12). Synchronous operation of turbine and pressure – regulating valve [10].

Solved Problems

Q.1). Estimate the energy dissipation rate in a cumulus cloud, both per unit mass and for the entire cloud. Base your estimates on velocity and length scales typical of cumulus clouds. Compute the total dissipation rate in Kilowatts. Also estimate the Kolmogorov micro scale η. Use $\rho = 1.25\ kg/m^3$ and $v = 15 \times 10^{-6}\ m^2/sec$.

Solution:

a) For typical cumulus, we can assume

$$l = 1\ km \quad u = 0.3\ m/sec \quad \text{and} \quad \frac{l}{L} = 0.2 \quad \therefore L = 5000\ m$$

∴ Energy dissipation rate/unit mass, is given by

$$\frac{\in}{kg} \sim \frac{u^3}{l}$$

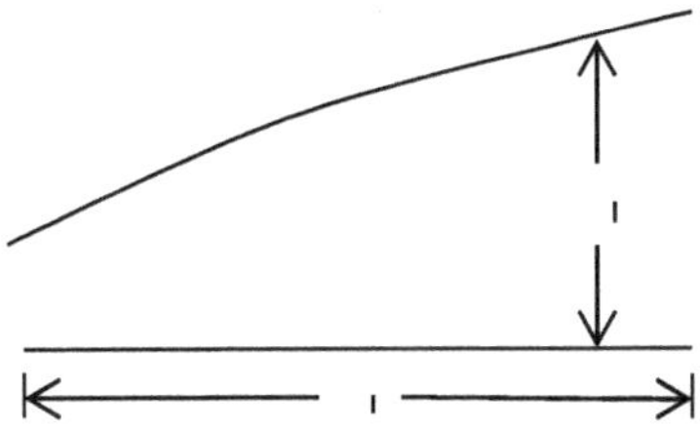

Energy dissipation rate (entire cloud) is given

$$\in \sim u^3/l\ .\ \text{mass}$$
$$\text{or} \quad \in \sim u^3/l\ .\ \rho L l.1 \quad \text{(unit depth)}$$

b) Total dissipation rate in Kilowatts is

$$\in \sim \frac{u^3}{l}\ .\ \rho L l.1$$

$$\in \sim \frac{0.3^3}{1000\ m} \cdot \frac{m^3}{s^3}\ 1.25 \cdot \frac{kg}{m^3} \cdot 1000\ m\ .5000\ m\ .1$$

$$\in \sim 168.75\ \frac{kg.m}{s^3}.m \sim 168.75 \cdot \frac{N.m}{s}$$

$$\in \sim 168.75.\frac{J}{s} \sim 168.75\ Watt$$

$\therefore \qquad \in \sim 0.16875$ Kilowatts.

c) Kolmogorov micro scale ŋ is given by

$$ŋ = \left(\frac{v^3}{\in/kg}\right)^{1/4} = \left((15 \times 10^{-6})^3 \frac{m^6}{s^3} \Big/ \left(0.3^3 \frac{m^6}{s^3} /1000\ m\right)\right)$$

$$ŋ = \frac{(15 \times 10^{-6})^3 \times 1000}{0.3^3} = 0.0033437\ m$$

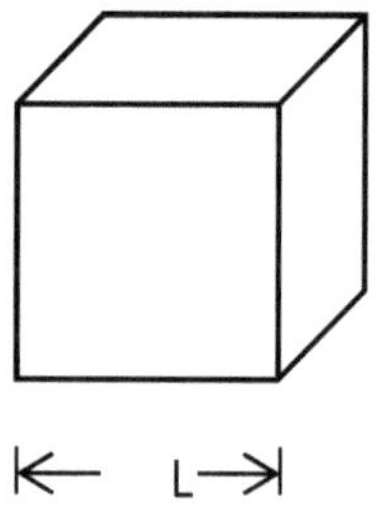

Q.2). Length scale = L

a) Derive an expression for decay of the kinetic energy $\frac{3}{2}\ u^2$ as a function of time.

b) Compute C by requiring that the dissipation rate is continuous at $uL/v = 10$? $i.e\ \in = cvu^2/L^2$.

c) Derive an expression for the decay of the K.E. when $uL/v < 10$?

d) If $L = 1\ m, v = 15 \times 10^{-6}\ m^2/s$ and $u = 1\ m/s,\ t = 0$, how long does it take before the turbulence enters the final viscous effect dominate the turbulent period of decay?

e) If the effect on the walls of the box on decay of the turbulence may be ignored. Can you support this assumption in any way?

Solution:

a) $K.E. = \frac{3}{2}\, u^2$

Decay is given by $\frac{d}{dt}\,(K.E.) \sim \frac{u^2}{T} \sim \frac{u^3}{l}$ where $T = \frac{L}{u}$

$$\therefore \quad \frac{d}{dt}\,(K.E.) = \frac{3}{2}\frac{d}{dt}\,(u^2)$$

$$= \frac{3}{2}\frac{d}{dt}\left(\left(\frac{L}{T}\right)^2\right) \quad \text{where } \frac{L}{T} = u$$

$$= \frac{3}{2}\,L^2\frac{d}{dt}\,(T^{-2})$$

$$\therefore \quad \frac{d}{dt}\,(K.E.) = -3\,\frac{L^2}{T^3}$$

b) Since dissipation rate is continuous it means the dissipation rate at $ul/v >$ 10 which is given $\in \approx u^3/l$ and dissipation rate at $ul/v < 1$ which is given by $\in \approx Cv\, u^2/L^2$ are

Hence

$$u^3/L = Cv\, u^2/L^2$$

$$C = \frac{uL}{v} = 10 \qquad \therefore \; C = 10$$

c) When $\frac{uL}{v} < 10,$ decay of the K.E. is given by

$$\frac{d}{dt}\left(\frac{3}{2}u^2\right) = 10\, v\, u^2/L^2 \quad \text{at} \quad \frac{uL}{v} < 10$$

where we have used the assumption that when $\frac{uL}{v} < 10$ and the dissipation is continous the rate of dissiption is govern by equation $10\,v\frac{u^2}{L^2}$

d) If $L = 1\,m, v = 15 \times 10^{-6}\ m^2/s$ Find T =?
rate of dissipation is given by

$$\epsilon = \frac{u^3}{L} \equiv 10\,v\frac{u^2}{L^2} \approx 10\,v\,\frac{1}{T^2}$$

$$T^2 = \frac{10\,vL}{u^3} = 150 \times 10^{-6}$$

$$T = 0.0122\,sec$$

e) From the given information we can say that the problem deals with small scale. As we know that for small scale, the viscous effects are important. Therefore, this case we can't neglect the effect of the wall of the box on the decay of the turbulence. On the other hand, if we consider the system as a large scale in this case we can neglect the wall effect because the viscous effect is important in this situation. To decide whether the system is small scale or large scale it needs a lot experience in turbulent field.

Q.3). Given

Large eddies

Length scale = ℓ
Velocity scale v (l) = u
Time scale t (l) = ℓ/u

Small eddies

Length scale η
Velocity scale v
Time scale τ

Required:

a) Estimate the characteristic velocity v (r) and characteristic time t (r) of eddies of size r, where r any length in the rang $\eta < r < \ell$. Do this by assuming that v (r) are determined by $\in$.

b) Show that your results agree with the known velocity and time scales at r = l and r = η.

Solution:

a) Large eddy intermediate eddy ~ small eddy

$$\in \sim \left(\frac{u^3}{l}\right)_{large} \sim \left(\frac{v^3}{r}\right)_{\text{intermediate}}$$

$$\therefore \quad \frac{v^3}{r} = \in \quad \therefore \quad v = (\in r)^{\frac{1}{3}} \qquad \textbf{(1)}$$

$$\text{and} \quad v\,(r) = \frac{r}{t(r)} \quad \Rightarrow \quad t(r) = \frac{r}{v\,(r)} = \frac{r}{(\in r)^{\frac{1}{3}}}$$

$$\therefore \quad t(r) = \left(\frac{r^2}{\in}\right)^{\frac{1}{3}} \qquad \textbf{(2)}$$

b) To check, for small scale $r = \eta = \left(\frac{\upsilon^3}{\in}\right)^{\frac{1}{4}}$ *and* $v = \upsilon$

From equation (1) $v = (\in r)^{\frac{1}{3}} = \left(\in \left(\frac{\upsilon^3}{\in}\right)^{\frac{1}{4}}\right)^{\frac{1}{3}}$

$$\therefore \quad v = \left[\frac{\in \upsilon^{\frac{3}{4}}}{\in^{\frac{1}{4}}}\right]^{\frac{1}{3}} = \left[\in^{\frac{3}{4}} \upsilon^{\frac{3}{4}}\right]^{\frac{1}{3}}$$

$\therefore \quad v = (\in \upsilon)^{\frac{1}{4}}$ which is given by eq. () for small scale.

Check time scale

$$r = \eta = \left(\frac{\upsilon^3}{\epsilon}\right)^{\frac{1}{4}} \quad \text{and} \quad t(r) = \tau$$

From eq. (2)

$$t(r) = \tau = \left(\frac{\eta}{\epsilon}\right)^{1/3} = \left(\left(\frac{\upsilon^3}{\epsilon}\right)^{\frac{1}{4}} \Big/ \epsilon\right)^{1/3}$$

$$\tau = \left(\upsilon^{3/2}/\epsilon^{1/2}.\epsilon\right)^{1/3} = \left(\upsilon^{3/2}/\epsilon^{3/2}\right)^{1/3}$$

$$\therefore \ \tau = \left(\frac{\upsilon}{\epsilon}\right)^{1/2}$$

For large scale

$$r = l \quad \text{and} \quad \mathrm{v} = u$$

From eq. (1)

$$\mathrm{v} = u = (\epsilon\ l)^{\frac{1}{3}}$$

Hence

$$u = (\epsilon\ l)^{\frac{1}{3}}$$

and from eq. (2)

$$t(r) = T_t = \left(\frac{l^2}{\epsilon}\right)^{1/3}$$

$$\therefore \quad T_t = \left(\frac{l^2}{\epsilon}\right)^{1/3}$$

$$T_t = \left(\frac{l^2}{u^3/l}\right)^{1/3} = \left(\frac{l^3}{u^3}\right)^{1/3} = l/u$$

$$\therefore\ T_t = \frac{l}{u}$$

Q.4). Given

$v = 50\,m/s,\quad u = 0.5,\quad l = 100\,m\,,\quad \upsilon = 15 \times 10^{-6}\ m^2/s$

Required

a) What is the highest frequency the anemometer will encounter?
b) What should the length of the hot- wire sensor be?
c) What is the permissible nose level?

Solution:

a) Relation between small and large scale

$$\frac{\tau}{t} \sim \frac{\left(\frac{\upsilon}{\epsilon}\right)^{\frac{1}{2}}}{\frac{l}{u}} = \frac{\left(\frac{\upsilon}{u^3/l}\right)^{\frac{1}{2}}}{\frac{l}{u}} = \left(\frac{ul}{\upsilon}\right)^{-\frac{1}{2}} = Re^{-\frac{1}{2}}$$

$$\therefore\ \tau = \frac{l}{u}.\left(\frac{ul}{\upsilon}\right)^{-\frac{1}{2}} = \frac{100}{0.5}\Big/\left(\frac{0.5 \times 100}{15 \times 10^{-6}}\right)^{-\frac{1}{2}}$$

$$\tau = \frac{200}{\sqrt{3.333 \times 10^{-6}}} = 0.1095\,\text{sec}$$

Frequncy $f = \frac{1}{\tau}$ $\quad f = 9.132\,Hz$

b) $$\frac{\eta}{l} = \left(\frac{ul}{\upsilon}\right)^{-\frac{3}{4}}$$

$$\eta = \frac{l}{\left(\frac{ul}{v}\right)^{\frac{3}{4}}} = \frac{100}{\left(\frac{0.5 \times 100}{15 \times 10^{-6}}\right)^{0.75}}$$

$$\eta = 1.2819 \times 10^{-3}\, m \sim 1.2819\, mm$$

c) Noise level which is given by the ratio of velocity in small scale to the velocity in large scale

Hence

$$\frac{\mathrm{v}}{u} = Re^{-\frac{1}{4}} = \frac{1}{\left(\frac{ul}{v}\right)^{0.25}} = \frac{1}{\left(\frac{0.5 \times 100}{15 \times 10^{-6}}\right)^{0.25}}$$

$\frac{\mathrm{v}}{u} = 0.0234 = 2.3$ % noise level should be less than this percentage.

Q.5). Discuss the mechanism of the motion in Prandtl mixing theory.

a- Discuss the origin of the transverse velocity fluctuation in mixing length theory.

Solution:

a) A lump or fluid comes from a layer at (y_l - l) and has velocity U (y_l - l) is displaced over a distance lm the transverse direction.

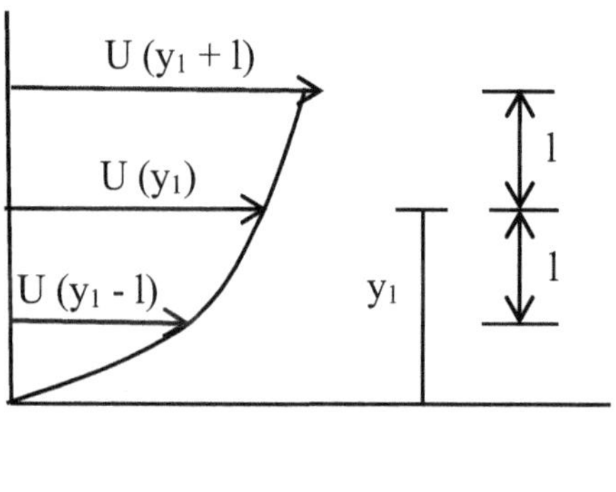

This distance is known on Prandtl mixing length its velocity in new lamia at y_1 is smaller than velocity prevail then the difference

$$\Delta u_1 = U(y_1) - U(y_1 - l) \approx l\left(\frac{dU}{dy}\right)_1$$

Similarly

$$\Delta u_2 = U(y_1 + l) - U(y_1) \approx l\left(\frac{dU}{dy}\right)_1$$

Time-average of the absolute value of fluctuation

$$\left|\overline{u'}\right| = \frac{1}{2}|(\Delta u_1) + (\Delta u_2)| \approx l\left(\frac{dU}{dy}\right)$$

b) Origin of $\bar{v}$

If two lumps appear in the reverse order they will move a part at a velocity $2u'$ and the empty space between them will befilled from the fluid, a gain rise to a transvers velocity compute v' two directions at y_1

$$\left|\overline{v'}\right| = \text{constant}\left|\overline{u'}\right| = \text{constant}\, l\left(\frac{dU}{dy}\right)_1$$

Q.6). Show that, for large Reynolds number:

a- Eddy diffusivity is much large than the molecular diffusivity.

b- Turbulent flow penetrates much deeper into the atmosphere than laminar flow.

Solution:

a) Diffusion eq. $\frac{\partial \theta}{\partial T} = k\frac{\partial^2 \theta}{\partial x_i \partial x_j}$

Time scale (theoretical) $T \sim \frac{L^2}{k}$

Time scale (actual) $T_1 \sim \frac{L}{u}$

Equate $\frac{L^2}{k} = \frac{L}{u} \quad \rightarrow k \sim uL$

Compare with kinematic viscosity

$$\frac{k}{\gamma} \sim \frac{k}{\upsilon} \sim \frac{uL}{\upsilon} \gg 1 \quad \text{to be turbulent} \quad \sim Re$$

Eddy diffusivity is much large than molecular diffusivity

b) Thickness of B.L (laminar)

$$L_m{}^2 \sim \upsilon T$$

Thickness of B.L (turbulent)

$$L_t \sim uT$$

The ratio will be $\frac{L_t}{L_m} \sim \frac{uT}{(\upsilon T)^{1/2}} \sim \left(\frac{u^2}{f\upsilon}\right)^{1/2} \sim \left(\frac{uL_t}{\upsilon}\right)^{1/2}$

$\frac{L_t}{L_m} = Re^{1/2}$ turbulent flow penetrates much deeper into atmospher than laminar

Q.7). For turbulent flow in channel

a- Determine the pressure variation P(y) in the direction normal to the flow by considering the momentum equation perpendicular to the wall.

b- Show that P is maximum at the wall.

c- The minimum P value is related to the maximum value of $\overline{\upsilon^2}$

Solution:

$$0 = -\frac{1}{\rho}\frac{\partial P}{\partial y} - d\,\overline{v'^2} \quad \Rightarrow \quad \frac{\partial P}{\partial y} = \rho\, d\,\overline{v'^2}$$

$$\text{at wall } \frac{\partial P}{\partial y} = \rho\, d\,\overline{v'^2} \quad \Rightarrow \quad \frac{\partial P}{\partial y} = 0$$

at P max

$$\frac{\partial P}{\partial y} \approx \left|v'^2\right| \text{ based on moment equation}$$

Q.8).

a- Let u' and v' be given by two since waves of equal amplitude and frequency. Determine the bounds on k_{uv} .

b- Consider a class of problem for which $l = ky^2$. Determine the form of the time-averaged turbulent velocity profile (U). In a region in which the turbulent shearing stress is assumed to be constant. Could this profile satisfy a wall boundary condition?

Solution:

a- Let $u' = \text{A sinwt}$ and $v' = \text{A sinwt}$

Then as $T \rightarrow$ large

$$u'v' = \frac{A^2}{T}\int_0^T sinwt^2\, d$$

$$= \frac{A^2}{T}\int_0^T \left(\frac{1}{2} - \frac{1}{2}\, cos2wt\right) dt = \frac{A^2}{2}$$

$$\therefore \sqrt{u'}\sqrt{v'} = \frac{A}{\sqrt{2}} \text{ .thus}$$

$$k_{uv} = \overline{u'v'}/\sqrt{\overline{u'^2}}\sqrt{\overline{v'^2}} = \frac{A^2/2}{A/\sqrt{2}.A/\sqrt{2}} = 1$$

Let $u' = \mathrm{A\,sinwt}$ and $v' = -\mathrm{A\,sinwt}$

Then $\overline{u'v'} = -\frac{A^2}{2}$ and $k_{uv} = -1.$

The above represents wares in phase and 180° out of phase.

For wares 90° out of phase $u' = \mathrm{A\,sinwt}$ and $v' = \mathrm{A\,sinwt}$ then $k_{uv} = 0.$

The bonds are $-1 \leq k_{uv} \leq 1$

b-
$$\tau_t = \rho l^2 \left(\frac{dU}{dy}\right)^2 = \rho k^2 y^4 \left(\frac{dU}{dy}\right)^2 = \text{constant}$$

$$\therefore \quad \frac{dU}{dy} = -\frac{c_1}{y} + c_2 \qquad \text{where } c_1 = \sqrt{\rho k^2} = \text{constant}$$

Assuming $U = U(y)$ the solution is

$$U = \frac{c_1}{y} + c_2 \qquad \text{velocity profile}$$

At wall $l = 0$

$\therefore$ The profile would not satisfy a wall condition since $U(0) = \infty$.

Q.9). Select the correct answer

1- In turbulent flow, viscous shear stress

a- Decreases the internal energy of fluid at expense of K.E of turbulence.

b- Increases the K.E of turbulence on expense of internal energy of fluid.

c- Equitizes the internal energy of fluid with K.E of turbulence.

d- Increases the internal energy of at expense of K.E of turbulence.

e- None of these.

Solution:

(1-b)

2- Turbulence depends on environment, therefore

a- It is a feature of the fluids.

b- Single theory for all kinds and types of flow in general.

c- It needs a vortex stretching of 2D flows.

d- No single theory for all kinds and types of flow in general.

e- None of these.

Solution:

(2-d)

3- Turbulence flow originates from

a- An instability of flow for all Re numbers.

b- Localized 2D turbulence which arise at flow.

c- Interaction of viscous terms with pressure terms.

d- None of these.

e- An instability of flow if the Re becomes too large.

Solution:

(3-e)

Q.10).

a- Show that "More than one characteristic length are needed to deal with the boundary layer problems (Laminar)".

b- Formulate the rate of energy supply to small scale and the rate of viscous energy loss.

Solution:

a) For laminar B.L

$$U_l \frac{\partial U_j}{\partial x_j} = -\frac{1}{\rho}\frac{\partial P}{\partial x_i} + \upsilon \frac{\partial^2 U_i}{\partial x_i \partial x_j}$$

Suppose we have only one characteristic length L and velocity U

$$\therefore \quad \frac{U^2}{L} \quad \sim \quad \upsilon \frac{U}{L^2}$$

Take the ratio of intertie/viscous

$$= \frac{U^2 L^2}{L \upsilon U} = \frac{UL}{\upsilon} = Re$$

For high Re viscous must be neglected, B.C^s. and I.C^s. make it possible

∴ For one scale means viscous term out, which is impossible (no. B.L)

∴ Define another length scale (say l)

Again

$$\frac{U^2}{L} \sim \upsilon\frac{U}{l^2} \Rightarrow \frac{U}{L} = \frac{\upsilon}{l^2}$$

$\therefore \frac{l}{L} = \left(\frac{\upsilon}{UL}\right)^{1/2} = \left(\frac{1}{Re}\right)^{1/2}$ where l is transverse length (thickness)

b) Rate of energy supply to small scale ∈

$$\in \sim \frac{u^3}{l}$$

and rate of viscous energy loss

$$\frac{\upsilon u^2}{l^2}$$

Q.11).

a- In shear flow, discuss the vortex stretching term (production term).

b- In shear flows determine the sign of correlation coefficient k_{uv}.

c- What kind of eddies that are more effective than most in maintaining the desired correlation between *u* and *v* and in extracting energy from the mean flow? Discuss.

Solution:

a) Vortex stretching productum

1- Amplify or attenuate the vorticty by the elongation of vortex tube element

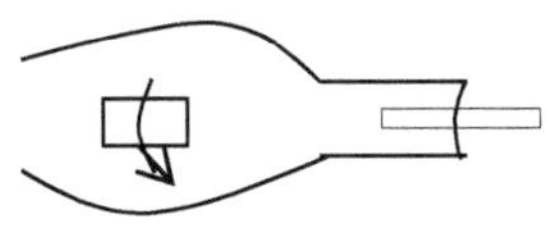

2- Produce x-component vorticty by reorientation the vortex tube which originally continued only in y-component vorticity.

b) at (A) $v' > 0$ $u' < 0$ $\therefore$ $u'v' < 0$

and $v' < 0$ $u' > 0$ $\therefore$ $u'v' < 0$

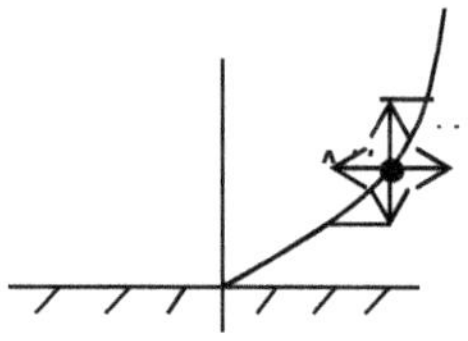

Flow in open

at (B) $v' > 0$ $u' > 0$ $\therefore$ $u'v' > 0$
and $v' < 0$ $u' < 0$ $\therefore$ $u'v' > 0$

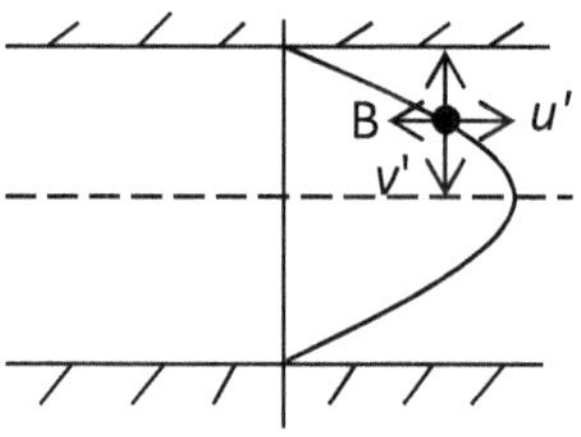

Flow in open pipes

at center line $v' > 0$ $u' = 0$ $u'v' = 0$

$v' < 0$ $u' = 0$ $u'v' = 0$

c) the most effective eddies are vortices whose principle axis is roughly aligned with that of the mean strain rate

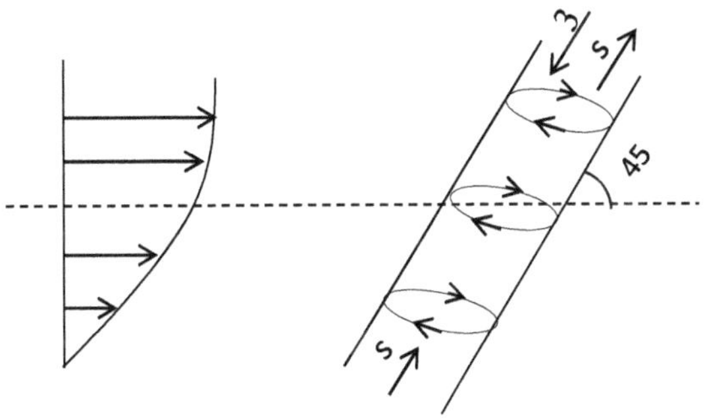

$$\overline{u_1{}'w_3{}'} \sim ul\frac{\partial \Omega_3}{\partial x_2}$$

Now compare of

$$\frac{\partial}{\partial x_2}\left(-\overline{u_1{}'u_2{}'}\right) = \overline{u_2{}'w_3{}'} - \overline{u_3{}'w_2{}'}$$

With

$$\frac{\partial}{\partial x_2}\left(-\overline{u_1{}'u_2{}'}\right) = ul\frac{\partial \Omega_3}{\partial x_2} - u\,\Omega_3\frac{\partial l}{dx_2}$$

We conclude that the negative of

$\overline{u_3{}'w_2{}'}$ is associated with a change of

Scale effect

$$\overline{u_3{}'w_2{}'} \sim u\,\Omega_3\frac{\partial l}{dx_2}$$

Q.12). K. Energy equation of motion for steady mean flow is given by: (with normal notations)

$$U_j\frac{\partial}{\partial x_j}\left(\frac{1}{2}U_iU_i\right) = \frac{\partial}{\partial x_j}\left(-\frac{P}{\rho}U_j + 2\,\upsilon u_i S_{ij} - \overline{u_i{}'u_j{}'}\,U_i\right) - 2\,\upsilon S_{ij}S_{ij} + \overline{u_i{}'u_j{}'}\,S_{ij}$$

Show that, the work done by turbulent parts will be much greater than the work done by viscous effect.

Solution:

$$U_j\frac{\partial}{\partial x_j}\left(\frac{1}{2}U_iU_i\right)=\frac{\partial}{\partial x_j}\left(-\frac{P}{\rho}+U_j+2\upsilon U_iS_{ij}-\overline{u_i'u_j'}\,U_i\right)-2\upsilon S_{ij}S_{ij}+\overline{u_i'u_j'}\,S_{ij}$$

2   3   4   1

Compare (1) and (2)

$$\overline{u_i'u_j'}\,S_{ij}\sim uuu/l\sim u^3/l$$

Hence $\overline{u_i'u_j'}\,S_{ij}=C_1\,\mathrm{u\,l}\,S_{ij}S_{ij}$ where $S_{ij}\sim\frac{u}{l}$

Compare with (1) we have

$$C_1\,\mathrm{u\,l}\,S_{ij}S_{ij}/2\,\upsilon S_{ij}S_{ij}=Re\gg 1$$

Now Compare (3) and (4) by the same way

$$C_2\,\mathrm{u\,l}\,S_{ij}S_{ij}/2\,\upsilon S_{ij}S_{ij}=Re\gg 1$$

∴ work done by turbulent parts will be much greater than the work done by viscous effect.

Q.13). For 2D flow, the equation of motion in U_1 direction is given by (neglecting the viscosity)

$$U_1\frac{\partial U_1}{\partial x_1}+U_2\frac{\partial U_1}{\partial x_2}=-\frac{1}{\rho}\frac{\partial P}{\partial x_i}+\overline{u_2'w_3'}-\overline{u_3'w_2'}$$

Compare with momentum eq.

$$U_i\frac{\partial U_j}{\partial x_j}=-\frac{1}{\rho}\frac{\partial P}{\partial x_i}+\upsilon\frac{\partial^2U_i}{\partial x_i\partial x_j}-\frac{\partial}{\partial x_j}\overline{u_i'u_j'}$$

And use the modeling of the terms to show that the nature of $\overline{u_3'w_2'}$ is associated with a change of scale effect.

Solution:

For 2D flow (for most B.L & wake flows)

$$U_1 \gg U_2 \quad U_3 = 0 \quad \frac{\partial}{\partial x_1} \ll \frac{\partial}{\partial x_2}$$

only nonzero component & Ω_i is $\Omega_3 = \frac{dU_2}{dx_1} - \frac{dU_1}{dx_2} = -\frac{dU_1}{dx_2}$

compare

$$\frac{\partial}{\partial x_2}\left(-\overline{u_1'u_2'}\right) = \overline{u_2'w_3'} - \overline{u_3'w_2'}$$

Model

$$\overline{u_1'u_2'} \sim ul\frac{\partial U_1}{\partial x_2} \rightarrow \frac{\partial}{\partial x_2}\left(-\overline{u_1'u_2'}\right) \sim ul\frac{\partial^2 U_1}{dx_2{}^2} + u\frac{\partial l}{dx_2}\frac{\partial U_1}{\partial x_2}$$

Where

$$\frac{\partial U_1}{\partial x_2} = -\Omega_3$$

$$\therefore \quad \frac{\partial}{\partial x_2}\left(-\overline{u_1'u_2'}\right) \sim ul\frac{\partial \Omega_3}{\partial x_2} - u\frac{\partial l}{dx_2}\Omega_3$$

Now, model $\overline{u_2'w_3'}$ associated with $\overline{u_3'u_2'}$

Q.14). Show that, for channel flow and $Re^* \rightarrow \infty$

a- $\left(-\frac{\overline{uv}}{u_x^2} + \frac{1}{Re^*}\frac{d(U/u^*)}{d(y/h)} = 1 - \frac{y}{h}\right)$, is not valid near the wall

b- $\left(-\frac{\overline{uv}}{u_x^2} + \frac{1}{Re^*}\frac{d}{d\eta}\left(\frac{U}{u^*}\right) = 1 - \eta\right)$, for outer-region, Reynolds stress is order of one

c- $\left(U_j\frac{\partial U_i}{\partial x_j}\right)$, are zero when the velocity profile not depends on downstream distance.

d- $\left(0 = -\frac{y}{\rho}\frac{dP_o}{dx} - \overline{uv} + v\frac{dU}{dy} - u^2 x\right)$, under what boundary conditions the shear stress at the wall is determined by the pressure and channel width.

Solution:

a) $-\frac{\overline{uv}}{u^{*2}}+\frac{1}{Re^*}\frac{d(U/u^*)}{d(y/h)}=1-\frac{y}{h}$

For $Re^* \rightarrow 0$, the viscous term is zero, at the wall, the stress is purely viscous

$\therefore$ the above eq. Is not valid at the wall

b) $-\frac{\overline{uv}}{u^{*2}}+\frac{1}{Re^*}\frac{d}{d\eta}\left(\frac{U}{u^*}\right)=1-\eta$

If $Re \rightarrow \infty$ $\quad\therefore \frac{\overline{uv}}{u_*^2}=1$ $\therefore$ Rey rated stress is order of one

c) $U_j\frac{\partial U_i}{\partial x_j}$ for fully developed flow interia terms are zero.

d) at center line where $\left(-\rho\,\overline{u'v'}+\mu\frac{dU}{dy}\right)=0$.

at $y=h$

$$\therefore\ u^{*2}=-\frac{h}{\rho}\frac{dp_o}{dx}$$

Q.15). In pipe flow shear stress varies linearly within distance from the wall $\tau=\tau_w\left(1-\frac{y}{R}\right)$, where R: is the pipe raduis. Find the variation of kinematic viscosity for turbulent flow and the mixing. Length with other parameters. (u^*, y, R)

Solution:

$$\tau=\tau_\circ\left(1-\frac{y}{R}\right)=\rho\in\frac{du}{dy} \qquad \textbf{(1)}$$

And for prandial mixing length

$$\tau = \rho\, l \left(\frac{du}{dy}\right)^2 = \rho\, k^2 y^2 \left(\frac{du}{dy}\right)^2$$

$$\frac{\tau/\rho}{k^2 y^2} = \frac{u^{*2}}{k^2 y^2} \quad \text{or} \quad \frac{u^*}{ky} = \frac{du}{dy} \qquad \textbf{(2)}$$

From 1,2 $\quad \tau = \tau_\circ\left(1 - \frac{y}{R}\right) = \rho \in . \frac{u^*}{ky} \qquad \tau_\circ/\rho = u^{*2}$

$$\therefore \in = \tau_\circ . ky . \frac{1}{\rho u^*}\left(1 - \frac{y}{R}\right)$$

∴ eddy viscosity:

$$\in = u^* ky \left(1 - \frac{y}{R}\right) \qquad \textbf{(3)}$$

Also from eq. 2 $\qquad \frac{u^*}{l} = \frac{du}{dy} \qquad$ **(4)**

From eq. (4.1)

$$\tau_\circ \left(1 - \frac{y}{R}\right) = \rho \in . \frac{u^*}{l}$$

$$\therefore \; l = \rho \in . u^* . \frac{1}{\tau_\circ \left(1 - \frac{y}{R}\right)}$$

$$l = \in . u^* . \frac{1}{u^{*2}} . \frac{1}{\tau_\circ \left(1 - \frac{y}{R}\right)} = \in . \frac{1}{u^* \left(1 - \frac{y}{R}\right)} \qquad \textbf{(5)}$$

Q.16). Water flow in 20 cm diameter pipe under fully developed conditions. The flow is turbulent and the centerline velocity 10 m/s. Compute the wall shear stress, the average velocity, the flow rate and the pressure drop. For length 100 m.

Where $= 0.4, B = 5, v = 1 \times 10^{-6}\, m^2/s$.

Solution:

$$\frac{u}{u^*} = \frac{1}{k}\ln\frac{y\,u^*}{\upsilon} + B \qquad \text{overlap law}$$

For flow in pipe $y = R - r$

At the center line $r = 0$

$$\therefore \quad \frac{u_{max}}{u^*} = \frac{1}{k}\ln\frac{R\,u_*}{\upsilon} + B$$

$$k = 0.4,\ \ B = 5,\ \ u_{max} = 10\frac{m}{s}, \qquad R = 0.1\,m,\ \ \upsilon = 1\times 10^{-6}\ m^2/s$$

$$\therefore \quad \frac{10}{u^*} = \frac{1}{0.4}\ln\frac{(0.1)\,u^*}{1\times 10^{-6}} + 5$$

$$\frac{10}{u^*} = \frac{1}{0.4}[\ln 100000 + \ln u^*] + 5$$

$$\frac{4}{u^*} = 11.5 + \ln u^* + 2$$

$$\frac{4}{u^*} = 13.5 + \ln u^*$$

or $4 = 13.5\,u^* + u^*\ln u^*$

$$u^* \approx 0.33\frac{m}{s}$$

$$\therefore\ \tau_w = \rho u^{*2} = 1000.(0.33)^2$$

$$= 109\ N/m^2$$

$$V_{averge} = u^*\left[2.44\ \ln\frac{Ru^*}{\upsilon} + 1.34\right]$$

$$= 0.33\left[2.44\ \ln\frac{0.1\times 0.33}{1\times 10^{-6}} + 1.34\right]$$

$= 8.82\ m/s$

$$Q = \text{AV} = \frac{\pi}{4}(0.2)^2 \times 8.82 = 0.277\ m^3/s$$

$$\Delta\text{P} = \frac{2 \times 109}{0.1} \times 100 = 218\ kPa.$$

Q.17). Prandtl developed a convenient exponential velocity distribution formula for turbulent flow in smooth pipe $\frac{u}{u_{max}} = \left(\frac{y}{R}\right)^{1/7}$, find and approximate expression for mixing length distribution $\frac{1}{R}$ as a function of y, R, f, where f is a function factor.

When the force balance for steady flow in pipe state. Shear stress $\tau = -\frac{r}{2}\frac{dp}{dl}$ and $\frac{\text{V}}{u_{max}} = \frac{1}{1+1.29\, f^{1/2}}$

Where $\frac{dp}{dl}$ the pressure drop and V the average velocity along the pipe.

Solution:

$$\frac{u}{u_{max}} = \left(\frac{y}{R}\right)^{\frac{1}{7}} \qquad \textbf{(1)}$$

$$\tau = -\frac{r}{2}\frac{dP}{dl} \qquad \tau_0 = -\frac{R}{2}\frac{dP}{dl}$$

$$\therefore\ \tau = \tau_0 \cdot \frac{r}{R} \qquad R - y = r$$

$$\therefore\ \tau = \tau_0\ \left(1 - \frac{y}{R}\right) = \rho l^2 \left(\frac{du}{dy}\right)^2$$

$$\therefore l = \frac{u^* \sqrt{1-y/R}}{du/dy} \qquad \textbf{(2)}$$

$$u^* = \sqrt{\frac{\tau_0}{\rho}}$$

Since $\frac{du}{dy} = u_{max} \cdot \frac{1}{7} \cdot \frac{1}{R} \left(\frac{y}{R}\right)^{-\frac{6}{7}}$ **(3)**

Substituting (3) in (2)

$$l = \frac{u^*}{u_{max}} .7R . \left(\frac{y}{R}\right)^{\frac{6}{7}} \sqrt{1 - \frac{y}{R}} \quad \textbf{(4)}$$

Since $\frac{u^*}{u_{max}} = \frac{V}{u_{max}} \times \frac{u^*}{V} = \frac{1}{1 + 1.29\sqrt{f}} \cdot \frac{\sqrt{f}}{\sqrt{8}}$

$$= \frac{\sqrt{f}}{2.83+3.65\sqrt{f}} \quad \textbf{(5)}$$

From eqs. (4 and 5)

$$l = 7R\left(\frac{y}{R}\right)^{\frac{6}{7}} \frac{\sqrt{f}}{2.83 + 3.65\sqrt{f}} \cdot \sqrt{1 - \frac{y}{R}}$$

Q.18). For oil flow in a pipe far downstream of the entrance, the axial velocity profile is a function of (r) only and is given by $u = \left(\frac{C}{\mu}\right)(R^2 - r^2)$ where C is a constant and R is the pipe radius. Suppose the pipe is (1 cm) in diameter and $u_{max} = 30\ m/s$. Compute the wall shear stress if $\mu = 0.3\ Pa.s$ and the discharge past a fixed cross-section.

Solution:

$$u = \frac{c}{\mu}(R^2 - r^2)$$

at $r = 0 \quad u = u_{max}$

$$\therefore u_{max} = \frac{c}{\mu} R^2 \qquad \therefore c = \frac{\mu}{R^2} u_{max}$$

$$\therefore u = \frac{\mu}{R^2} u_{max} \cdot \frac{1}{\mu}(R^2 - r^2)$$

$$u = u_{max}\left(1 - \left(\frac{r}{R}\right)^2\right)$$

$$\tau = \mu \frac{du}{dy} = -\mu \frac{du}{dr} \qquad y = R - r \qquad \therefore dy = -dr$$

$$\therefore \tau = -\mu \frac{d}{dr}\left(u_{max}(1 - \left(\frac{r}{R}\right)^2\right)$$

$$\tau = -\mu\, u_{max}\left[0 - \frac{2r}{R^2}\right] = \mu \frac{2r}{R^2} u_{max}$$

at the wall $r = R \quad \therefore \tau = \mu \frac{2R}{R^2} u_{max}$

$$\therefore \tau = \frac{2\mu}{R} u_{max} = \frac{2 \times 0.3 \times 30}{0.5 \times 10^{-2}} = 3600\ Pa.$$

$$Q = AV = \int_0^R u 2\pi r dr$$

$$\therefore V = \frac{2\pi}{\pi R^2} \int_0^R u_{max}(1 - \left(\frac{r}{R}\right)^2 r dr$$

$$= \frac{2\, u_{max}}{R^2}\left[\frac{r^2}{2} - \frac{r^4}{4R^2}\right]_0^R = \frac{2\, u_{max}}{R^2}\left[\frac{R^2}{2} - \frac{R^4}{4R^2}\right]$$

$$= \frac{2\, u_{max}}{R^2}\left[\frac{R^2}{2} - \frac{R^4}{4R^2}\right] = \frac{2\, u_{max}}{R^2} \times \frac{R^2}{2} = \frac{1}{2} u_{max}$$

$$V = \frac{1}{2}\, u_{max} = 15\ m/s$$

$$\therefore Q = \pi R^2 V = \frac{\pi(D^2)}{4} V$$

$$= \frac{\pi(1 \times 10^{-2})^2}{4} \times 15 = \frac{\pi}{4 \times 10000} \times 15$$

$$= 11.8 \times 10^{-4}\ m^3/s = 1.18\ l/s$$

Q.19). The relation for turbulent the shear stress express that:

$$\tau = \mu_t \frac{\partial u}{\partial y}$$

where μ_t = is the eddy viscosity for cootte flow between two parallel plates. Show that the eddy viscosity is given by:

$$\mu_t = k\rho u^* y$$

For $y = \leq \frac{h}{2}$ use the velocity gradient of over-lap eq. compute the ratio of:

$$\frac{\mu_t}{\mu} \text{ at } y = \frac{h}{2} \text{ if } Re_h = 10^5, f = 0.018 \text{ and } k = 0.41$$

Solution:

Overlap eq.

$$u = u^* \left[\frac{1}{k} \ln \frac{yu^*}{v} + B\right]$$

$$\therefore \frac{du}{dy} = \frac{u^*}{ky}$$

$$\text{Since } \tau = \mu_t \frac{du}{dy} = \mu_t \frac{u^*}{ky}$$

$$\therefore \ \mu_t = \frac{ky\tau}{u^*} \times \frac{u^*}{u^*} = \frac{ky\tau u^*}{u^{*2}}$$

Since $u^{*2} = \frac{\tau}{\rho}$

$$\therefore \ \mu_t = \rho k y u^*$$

Since τ is linear across the parallel plate:

And

$$Re_h = \frac{hV}{v}$$

And

$$\frac{\mu_t}{\mu} = \frac{\rho k y u^*}{\mu} = \frac{1}{2}\frac{\rho k h u^*}{\mu}\frac{V}{V} \qquad \leftrightarrow \quad y = \frac{h}{2}$$

$$\frac{\mu_t}{\mu} = \frac{1}{2}k\frac{u^*}{V} \times Re_h$$

and $v = \frac{\mu}{\rho}$ $\qquad \frac{u^*}{V} = \sqrt{\frac{f}{8}}$

$$\frac{\mu_t}{\mu} = \frac{1}{2}k\sqrt{\frac{f}{8}}\,Re_h = \frac{1}{2} \times 0.41\sqrt{\frac{0.018}{8}} \times 10^5$$

$$\frac{\mu_t}{\mu} = 972.4$$

Q.20). Show that the velocity ratio in turbulent flow in pipes states

$$\frac{v}{u_{max}} = \left(1 + 1.29\sqrt{f}\right)^{-1}$$

For $K = 0.41, f =$ friction factor, $v =$ average velocity and $u_{max} =$ maximum velocity

Solution:

Since $\frac{V}{u^*} = \frac{1}{k}\ln\frac{Ru^*}{v} + B - \frac{3}{2k}$ flow in pipe **(1)**

and at $r - 0$ $u = u_{max}$ for $y = R - r$

$\therefore \frac{u_{max}}{u^*} = \frac{1}{k}\ln\frac{Ru^*}{v} + B$ **(2)**

From eqs. (1, 2)

$$\frac{u_{max}}{u^*} - \frac{V}{u^*} = \frac{3}{2k} = \frac{3}{0.82}$$

$$\frac{u_{max}}{u^*} = \frac{V}{u^*} + \frac{3}{0.82}$$

$$\text{Since } \frac{V}{u^*} = \left(\frac{8}{f}\right)^{\frac{1}{2}}$$

$$\therefore \frac{u_{max}}{u^*} = \left(\frac{8}{f}\right)^{\frac{1}{2}} + \frac{3}{0.82}$$

Multiply both side by $\frac{V}{V}$

$$\frac{u_{max}}{u^*} \times \frac{V}{V} = \left(\frac{8}{f}\right)^{\frac{1}{2}} + \frac{3}{0.82}$$

$$\frac{u_{max}}{V} \times \frac{V}{u^*} = \left(\frac{8}{f}\right)^{\frac{1}{2}} + \frac{3}{0.82}$$

$$\frac{u_{max}}{V} \times \left(\frac{8}{f}\right)^{\frac{1}{2}} = \left(\frac{8}{f}\right)^{\frac{1}{2}} + \frac{3}{0.82}$$

$$\frac{u_{max}}{V} = 1 + \frac{3}{0.82} \cdot \frac{\sqrt{f}}{\sqrt{8}} = 1 + 1.29\sqrt{f}$$

$$\frac{V}{u_{max}} = \left(1 + 1.29\sqrt{f}\right)^{-1}$$

Q.21). For turbulent flow in pipe show that the average velocity of flow as follows:

$$V = u^* \left(\frac{1}{k}\ln\frac{Ru^*}{v} + B - \frac{3}{2k}\right)$$

Solution:

$$V = \frac{Q}{A} = \frac{1}{\pi R^2}\int_0^R u\ 2\pi r\, dr$$

$$y = R - r \quad \Rightarrow \ r = R - y \ \Rightarrow \ dr = -dy$$

at $r = 0 \ \Rightarrow \ y = R$ and at $r = R \ \ y = 0$

$$\therefore \ V = \frac{1}{\pi R^2}\int_R^0 u^* \left(\frac{1}{k}\ln\frac{yu^*}{v} + B\right) 2\pi(R - y)(-dy)$$

Or

$$V = \frac{2\pi u^*}{\pi R^2}\int_0^R \left[\frac{R - y}{k}\ln\frac{yu^*}{v} + B(R - y)\right] dy$$

Solved by parts

$$\int_0^R \frac{(R - y)}{k}\ln\frac{yu^*}{v}\, dy = uV - \int V\, du$$

$$u = \ln\frac{yu^*}{v} \quad \Rightarrow \quad du = \frac{dy}{y}$$

$$dV = \frac{R-y}{k}dy \quad \Rightarrow \quad dV = \frac{R}{k}dy - \frac{y}{k}\,\mathrm{dy}$$

$$\therefore \; V = \frac{R}{k}y - \frac{y^2}{2k}$$

$$\therefore \int_0^R \frac{(R-y)}{k}\ln\frac{yu^*}{v}\,dy = \left[\frac{R}{k}y\ln\frac{yu^*}{v} - \frac{y^2}{2k}\ln\frac{yu^*}{v}\right]_0^R - \int_0^R \left(\frac{R}{k} - \frac{y}{2k}\right)\mathrm{dy}$$

$$= \frac{R}{k}y\ln\frac{yu^*}{v} - \frac{y^2}{2k}\ln\frac{yu^*}{v} - \frac{R}{k}y + \frac{y^2}{4k}\Big]_0^R$$

$$= \frac{R^2}{k}\ln\frac{Ru^*}{v} - \frac{R^2}{2k}\ln\frac{Ru^*}{v} - \frac{R^2}{k} + \frac{R^2}{4k}$$

$$= \frac{R^2}{2k}\ln\frac{Ru^*}{v} - \frac{3R^2}{4k} \quad (1)$$

And

$$\int_0^R B\,(R-y\,)dy = BRy - \frac{By^2}{2}\Big]_0^R$$

$$= BR^2 - \frac{By^2}{2} = \frac{BR^2}{2} \quad \textbf{(2)}$$

From eqs. (1, 2)

$$V = \frac{2u^*}{R^2}\left[\frac{R^2}{2k}\ln\frac{Ru^*}{v} - \frac{3R^2}{4k} + \frac{BR^2}{2}\right]$$

$$= \frac{2u^*R^2}{R^2}\left[\frac{1}{2k}\ln\frac{Ru^*}{v} - \frac{3}{4k} + \frac{B}{2}\right]$$

$$V = u^*\left[\frac{1}{2k}\ln\frac{Ru^*}{v} - \frac{3}{2k} + B\right]$$

Q.22). Show that the above equation is accurate when the local velocity u equals the average velocity at $r = 0.777\ R$, regardless of Reynolds number.

Solution:

$$\text{Since}\ \frac{u_{(r)}}{u^*} = \frac{1}{k}\ln\frac{(R-\text{r})u^*}{v} + B$$

Where

$u_{(r)} =$ local velocity $\quad$ $\text{k} = 0.41$ $\quad$ $\text{B} = 5$

$\therefore\ u_{(r)} = V$

$$\therefore\ u^*\left[\frac{1}{k}\ln\frac{(R-\text{r})u^*}{v} + B\right] = u^*\left[\frac{1}{k}\ln\frac{Ru^*}{v} - \frac{3}{2k} + B\right]$$

$$\frac{1}{k}\ln\frac{(R-\text{r})u^*}{v} + B = \frac{1}{k}\ln\frac{Ru^*}{v} - \frac{3}{2k} + B$$

$$\ln\frac{Ru^*}{v} - \ln\frac{(R-\text{r})u^*}{v} = \frac{3}{2}$$

$$\ln\frac{\frac{Ru^*}{v}}{\frac{(R-\text{r})u^*}{v}} = \frac{3}{2}$$

$$\ln\frac{R}{(R-\text{r})} = \frac{3}{2}$$

$$\therefore\ \frac{R}{(R-\text{r})} = e^{1.5} = 4.48$$

$$\frac{R - \mathrm{r}}{R} = 0.223$$

$$1 - \frac{\mathrm{r}}{R} = 0.223$$

or $\frac{\mathrm{r}}{R} = 1 - 0.223$

$\therefore\ r = 0.777\,R$

Q.23). Prove that the stress tenser is symmetric

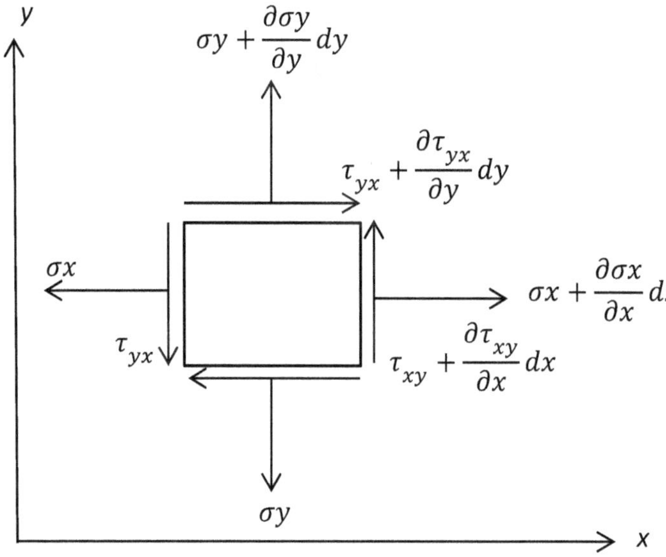

Write moment & momentum around (A)

$\sum(\vec{r}.\vec{f}) = \frac{D}{Dt}$ (Moment & momentum)

$$L.H.S = \left(\frac{\partial \sigma x}{\partial x}\delta_x.\delta_y \times 1\right)\frac{\delta_y}{2} - \left(\frac{\partial \sigma y}{\partial y}\delta_y.\delta_x \times 1\right)\frac{\delta_x}{2} + \left(\tau_{xy}.\delta_y.1\right)\delta_x - \left(\tau_{yx}.\delta_x.1\right)\delta_y$$

$$R.H.S = \frac{D}{Dt}(m.v.\ \text{dist})\rho\ \delta_x\delta_y \times 1$$

$$\rho\ \delta_x\delta_y \times 1 \qquad \delta_x \text{ or } \delta_y$$

Divide by $\delta_x\ \delta_y$. them take $\delta_x \rightarrow 0$ & $\delta_y \rightarrow 0$ to get a point

$$\tau_{xy} - \tau_{yx} = 0$$

$$\tau_{xy} = \tau_{yx}$$

$$i.e\ \tau_{ij} = \tau_{ji}$$

Q.24). Ethyl alcohol, $\rho = 789\ kg/m^3, \upsilon = 1.5 \times 10^{-6}\ m^2/s$ flows at $V = 300\ cm/s$ through 10 cm diameter drawn tubing-compute (a) the head loss per $100\ m$ of tube (b) the wall shear stress, and (c) the local velocity u at $r = 2$ cm By what percentage is the head loss increased due to the roughness of the tube ?

Solution:

$$Re_d = \frac{Vd}{\upsilon} = \frac{300 \times 10 \times 10^{-4}}{1.5 \times 10^{-6}} = 200\ 000$$

$$\because\ Re_d > 10^5$$

$$\therefore f = \left(1.8 \log \frac{Re_d}{6.9}\right)^{-2} = 0.0155 \quad (f = 0.01563 \text{ from eq.} (10))$$

$$(a)\ \ h_f = \frac{\Delta P}{\rho g} = f\frac{L}{d}.\frac{V^2}{2g} = 0.0155\ \frac{100}{0.1}\frac{(3)^2}{2 \times 9.81}$$

$$\Rightarrow h_f = 7.1105\ m$$

$$b)\ \frac{V}{u^*} = \left(\frac{\rho V^2}{\tau_w}\right)^{\frac{1}{2}} = \left(\frac{8}{f}\right)^{\frac{1}{2}}$$

$$\Rightarrow \tau_w = \frac{8}{f}\,\rho V^2 = \frac{0.0155}{8} \times 789 \times (3)^2$$

$$\therefore\ \tau_w = 13.758\ N/m^2$$

$$c)\ u_{(r)} = u^*\frac{1}{k}\ln\frac{(R-r)u^*}{v} + B$$

$$k = 0.41, \qquad B = 5.0, \qquad u^* = \sqrt{\frac{\tau_w}{\rho}} = \sqrt{\frac{13.758}{789}} = 0.132\ m/s$$

$$R - r = 5 - 2 = 3\ cm = 0.03\ m$$

$$\therefore\ u_{(r=2\ cm)} = \frac{0.132}{0.41}\ln\left(\frac{0.03 \times 0.132}{1.5 \times 10^{-6}}\right) + 5 \times 0.132$$

$$\therefore\ u_{(r)} = 3.1978\ m/s$$

Q.25). Air flow through a 14 cm tube under fully developed conditions. The centerline velocity is $u_0 = 5\ m/s$ and air density is $v_{air} = 1.5 \times 10^{-5}\ m^2/s$. Estimate from figure: $\rho_{air} = 1.2\ kg/m^3$

a) The friction velocity u^*.

b) The wall shear stress τ_w .

c) The average velocity V.

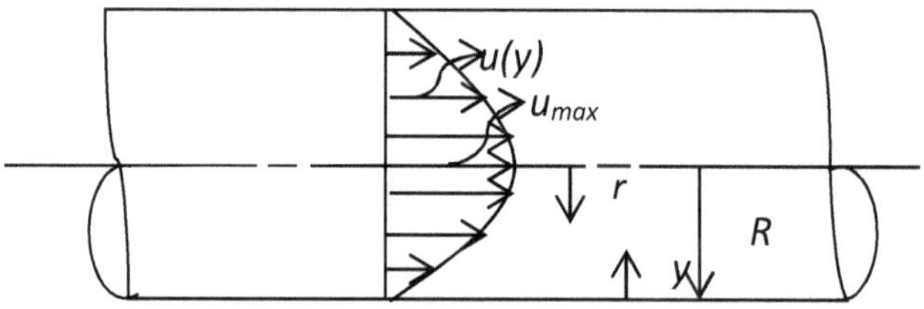

Solution:

$$\frac{u}{u^*} = \frac{1}{k}\ln\frac{yu^*}{v} + B$$

Logarithmic Law

a) $y =$ distance from wall $= R - r$

At the center $u = u_0,\ y = R$, the above eq. becomes,

$$\frac{u_0}{u^*} = \frac{1}{k}\ln\frac{Ru^*}{v} + B$$

Introducing $k = 0.41$ and $B = 5.0$ we obtain

$$\frac{u_0}{u^*} = \frac{1}{0.41}\ln\frac{Ru^*}{v} + 5.0$$

Where $u_0 = 5\ m/s\,, R = 0.07\ m,$ and $v_{air} = 1.5 \times 10^{-5}\ m^2/s$

$\therefore\ u^* = 0.228\ m/s$ (by trial and error) ans.(a)

b) Assuming the pressure and temp. Of air are 1 bar and 20 ^{0}C, respectively.

$$\rho_{air} = \frac{P}{Rt} = \frac{101325\ N/m^2}{287\ \frac{N.m}{kg.k}(20 + 273)k}$$

$$\Rightarrow \rho_{air} = 1.205\ \frac{kg}{m^3}$$

$$\text{Since } u^* = \sqrt{\frac{\tau_w}{\rho}} \Rightarrow \tau_w = \rho u^{*2} = 1.205(0.228)^2$$

$$\therefore \tau_w = 0.0626\ Pa \qquad \text{ans. (b)}$$

c) The average velocity

$$V = \frac{Q}{A} = \frac{1}{\pi R^2}\int_0^R U \times 2\pi r \, dr$$

$$\because \; u = u^*\left[\frac{1}{k}\ln\frac{yu^*}{v} + B\right], \qquad \text{and} \quad y = R - r$$

$$\Rightarrow \; u = u^*\left[\frac{1}{k}\ln\frac{yu^*}{v} + B\right] = u^*\left[\frac{1}{k}\ln\frac{(R-r)u^*}{v} + B\right]$$

$$\therefore V = \frac{2\pi u^*}{\pi R^2}\int_0^R \left[\frac{1}{k}\ln\frac{(R-r)u^*}{v} + B\right] r \, dr$$

$$\Rightarrow V = \frac{u^*}{2}\left(\frac{2}{k}\ln\frac{Ru^*}{v} + 2B - \frac{3}{k}\right) \quad \cdots\cdots (*) \text{ (proving it)}$$

$$= \frac{0.228}{2}\left(\frac{2}{0.41}\ln\left(\frac{0.07 \times 0.228}{1.51 \times 10^{-5}}\right) + 2 \times 5 - \frac{3}{0.41}\right)$$

$$\therefore V = 4.178 \; m/s \qquad \text{ans. (c)}$$

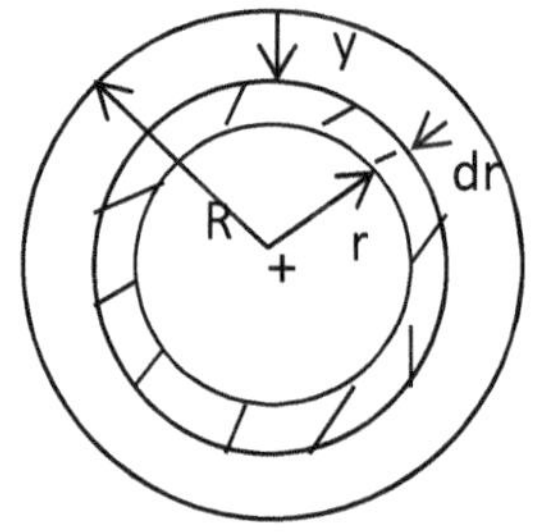

Proving Eq. (*) :

$$V = \frac{Q}{A} = \frac{1}{\pi R^2}\int_0^R U \times 2\pi r \, dr$$

$$\because y = R - r \quad \Rightarrow \quad R - y \quad \Rightarrow \quad dr = -dy$$

at $r = 0 \ \Rightarrow \ y = R$ and at $r = R \Rightarrow y = 0$

$$\therefore V = \frac{1}{\pi R^2} \int_0^R U^* \left(\frac{1}{k} \ln \frac{yu^*}{v} + B \right) \times 2\pi (R - y)\,(-dy)$$

$$= \frac{2\pi U^*}{\pi R^2} \int_0^R \left[\frac{(R - y)}{k} \ln \frac{yu^*}{v} + B(R - y) \right] dy$$

$$= BRy - \frac{By^2}{2}\bigg|_0^R = BR^2 - \frac{BR^2}{2} = \frac{BR^2}{2}$$

$$\int_0^R \frac{(R - y)}{k} \ln \frac{yu^*}{v}\, dy = uv - \int v\, du \qquad \text{Solved by parts}$$

$$u = \ln \frac{yu^*}{v} \ \Rightarrow du = \frac{dy}{y}$$

$$dv = \frac{(R - y)}{k} dy \ \Rightarrow dv = \frac{R}{y}\, dy - \frac{y}{k}\, dy \Rightarrow v = \frac{R}{k} y - \frac{y^2}{2k}$$

$$\therefore \int_0^R \frac{(R - y)}{k} \ln \frac{yu^*}{v}\, dy = \frac{R}{k} y \ln \frac{yu^*}{v} - \frac{y^2}{2k} \ln \frac{yu^*}{v} \bigg]_0^R - \int_0^R \left(\frac{R}{k} - \frac{y}{2k} \right) dy$$

$$= \frac{R}{k} y \ln \frac{yu^*}{v} - \frac{y^2}{2k} \ln \frac{yu^*}{v} - \frac{R}{k} y + \frac{y^2}{2k}\bigg|_0^R$$

$$= \frac{R^2}{k}\ln\frac{Ru^*}{v} - \frac{R^2}{2k}\ln\frac{Ru^*}{v} - \frac{R^2}{k} + \frac{R^2}{2k}$$

$$= \frac{1}{k}R^2\ln\frac{Ru^*}{v} - \frac{3}{4}\frac{R^2}{2k}$$

$$\therefore V = \frac{2u^*}{R^2}\left[\frac{R^2}{2k}\ln\frac{Ru^*}{v} - \frac{3}{4}\frac{R^2}{k} + \frac{3R^2}{2}\right]$$

$$V = \frac{u^*}{2}\left[\frac{2}{k}\ln\frac{Ru^*}{v} - \frac{3}{k} + 2B\right]$$

We should check the Reynolds number to ensure turbulent flow

$$Re_d = \frac{Vd}{v} = \frac{4.17 \times 0.14}{1.51 \times 10^{-5}} = 38700 > 4000$$

$\therefore$ The flow is definitely turbulent.

Q.26). For turbulent flow between two parallel plates, show that

$$\frac{1}{f^{1/2}} \simeq 2.0\ \log\left(Re_{Dh}f^{1/2}\right) - 1.19$$

Where D_h = hydraulic diameter = $2h$

f = friction factor

Using the logarithm law which state:

$$\frac{u_{(y)}}{u^*} = \frac{1}{k}\ln\frac{yu^*}{v} + B \qquad 0 < y < \frac{h}{2}$$

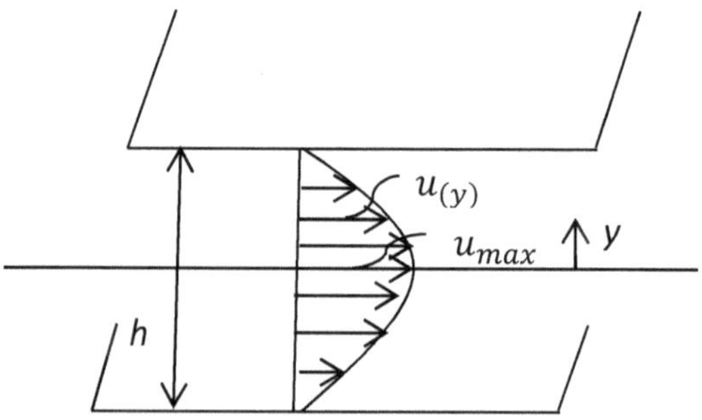

y

h

and $\dfrac{V}{u^*} = \left(\dfrac{8}{f}\right)^{1/2}$

Solution:

Assuming 1-unit depth

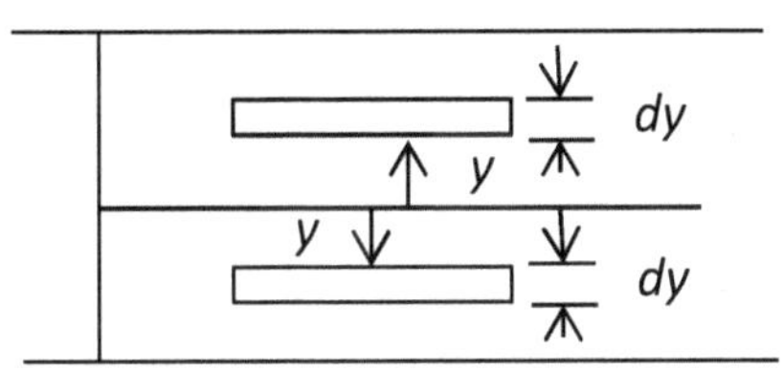

$$V = \frac{Q}{A} = \frac{1}{h}\int_0^{\frac{h}{2}} u^*\left(\frac{1}{k}\ln\frac{yu^*}{v} + B\right) \times 2dy$$

$$= \frac{2u^*}{h}\left[\int_0^{\frac{h}{2}} \frac{1}{k}\ln\frac{yu^*}{v}\, dy + \int_0^{\frac{h}{2}} Bdy\right]$$

$$= \frac{2u^*}{h}\left[\frac{y}{k}\ln\frac{yu^*}{v} - \frac{y}{k} + By\right]_0^{\frac{h}{2}}$$

$$= \frac{2u^*}{h}\left[\frac{h}{2k}\ln\frac{hu^*}{2v} - \frac{h}{2k} + \frac{Bh}{2}\right]$$

$$= u^*\left[\frac{1}{k}\ln\frac{hu^*}{2v} - \frac{1}{k} + B\right]$$

$$\Rightarrow \frac{V}{u^*} = \left(\frac{1}{0.41}\ln\frac{hu^*}{2v} - \frac{1}{0.41} + 5\right) \qquad (k = 0.41 \text{ and } B = 5.0)$$

$$\therefore \frac{V}{u^*} = 2.44\ln\frac{hu^*}{2v} + 2.56 \qquad \cdots\cdots(*)$$

$$\text{and } \frac{V}{u^*} = \left(\frac{8}{f}\right)^{1/2} \qquad \cdots\cdots(**) \text{ given}$$

$$\frac{hu^*}{2v} = \frac{2 \times V.hu^*}{2 \times 2.v.V} = \frac{1}{4}Re_{Dh}f^{1/2} \qquad \cdots\cdots(***)$$

$$\text{where } Re_{Dh} = \frac{2.V.hu^*}{v}$$

sub. Eq. $(**)$ and $(***)$ in to eq. $(*)$ gives,

$$\left(\frac{8}{f}\right)^{\frac{1}{2}} = 2.44\ln\left[\frac{1}{4}Re_{Dh}\left(\frac{f}{8}\right)^{\frac{1}{2}}\right] + 2.56$$

$$\frac{1}{\sqrt{f}} = 0.86267\ln\left[\frac{1}{4}Re_{Dh}\left(\frac{f}{8}\right)^{\frac{1}{2}}\right] + 0.905$$

$$= 0.86267\ln\left[\left(Re_{Dh}f^{\frac{1}{2}}\right) + \ln\left(\frac{1}{4}\right) - \ln(\sqrt{8})\right] + 0.905$$

$$= 0.86267\ln\left(Re_{Dh}f^{\frac{1}{2}}\right) - 2.09489 + 0.905$$

$$= 0.86267\ln\left(Re_{Dh}f^{\frac{1}{2}}\right) - 1.1876$$

$$\therefore \frac{1}{f^{\frac{1}{2}}} = 1.986\log\left(Re_{Dh}f^{\frac{1}{2}}\right) - 1.1876$$

$$\therefore \frac{1}{f^{\frac{1}{2}}} \simeq 2.0 \log\left(Re_{Dh} f^{\frac{1}{2}}\right) - 1.19$$

Q.27). Water flows in 20 cm diameter pipe under fully developed conditions. The flow is turbulent and the centerline velocity 10 m/s. Compute wall shear stress, the average velocity, the flow rate and the pressure drop. For length 100 m. Where $= 0.4, B = 5, v = 1 \times 10^{-6}\ m^2/s$.

Solution:

$$\frac{u}{u^*} = \frac{1}{k} \ln \frac{yu^*}{v} + B$$

For flow in pipe $y = R - r$

At the center line $r = 0$

$$\therefore \frac{u_{max}}{u^*} = \frac{1}{k} \ln \frac{Ru^*}{v} + B$$

$k = 0.4$ $B = 10\, m/s$ $R = 0.1\, m$ $v = 1 \times 10^{-6}$

$$\therefore \frac{10}{u^*} = \frac{1}{0.4} \ln \frac{(0.1)u^*}{1 \times 10^{-6}} + 5$$

$$\frac{10}{u^*} = \frac{1}{0.4} [\ln 100000] \ln u^* + 5$$

$$\frac{4}{u^*} = 11.5 + \ln u^* + 2$$

$$\frac{4}{u^*} = 13.5 + \ln u^*$$

$$\text{or } 4 = 13.5\, u^* + u^* \ln u^*$$

$$u^* \approx 0.33\ m/s$$

$$\therefore \tau_w = \rho u^{*2} = 1000\,.(0.33)^2$$

$$= 109\ N/m^2$$

$$V_{averge} = u^* \left[2.44 \ln \frac{Ru^*}{v} + 1.34\right]$$

$$= 0.33 \left[2.44 \ln \frac{0.1 \times 0.33}{1 \times 10^{-6}} + 1.34\right]$$

$$= 8.82\ m/s$$

$$Q = AV = \frac{\pi}{4}(0.2)^2 \times 8.82 = 0.277\ m^3/s$$

$$\Delta P = \frac{2 \times 109}{0.1} \times 100 = 218\ KPa.$$

Q.28). Prandtl developed a convenient exponential velocity distribution formula for turbulent flow in smooth pipe

$$\frac{u}{u_{max}} = \left(\frac{y}{R}\right)^{1/7}$$

Find and approximate expression for mixing length distribution $\frac{l}{R}$ as a function or y, R, f where f is a friction factor .

Shear stress $\tau = -\frac{r}{2}\frac{dp}{dl}$ and $\frac{v}{u_{max}} = \frac{1}{1+1.29\, f^{1/2}}$

Where $\frac{dp}{dl}$ the pressure drop and V the average velocity along the pipe .

Solution:

$$\frac{u}{u_{max}} = \left(\frac{y}{R}\right)^{\frac{1}{7}} \qquad \textbf{(1)}$$

$$\tau = -\frac{r}{2}\frac{dP}{dl} \qquad \tau_0 = -\frac{R}{2}\frac{dP}{dl}$$

$$\therefore \tau = \tau_0 \frac{r}{R} \qquad R - y = r$$

$$\therefore \tau = \tau_0 \left(1 - \frac{y}{R}\right) = \rho l^2 \left(\frac{du}{dy}\right)^2$$

$$\therefore l = \frac{u^* \sqrt{1-\frac{y}{R}}}{du/dy} \qquad \textbf{(2)}$$

$$u^* = \sqrt{\frac{\tau_0}{\rho}}$$

$$\text{Since } \frac{du}{dy} = u_{max} \cdot \frac{1}{7} \cdot \frac{1}{R} \left(\frac{y}{R}\right)^{-\frac{6}{7}} \qquad \textbf{(3)}$$

Substituting (3) in (2)

$$l = \frac{u}{u_{max}} \cdot 7R \cdot \left(\frac{y}{R}\right)^{\frac{6}{7}} \sqrt{1 - \frac{y}{R}} \qquad \textbf{(4)}$$

$$\text{Since } \frac{u}{u_{max}} = \frac{V}{u_{max}} \frac{u^*}{V} = \frac{1}{1 + 1.29\sqrt{f}} \cdot \frac{\sqrt{f}}{\sqrt{8}}$$

$$= \frac{\sqrt{f}}{2.83 + 3.65\sqrt{f}} \qquad \textbf{(5)}$$

From (4) , (5)

$$l = 7R \cdot \left(\frac{y}{R}\right)^{\frac{6}{7}} \cdot \frac{\sqrt{f}}{2.83 + 3.65\sqrt{f}} \cdot \sqrt{1 - \frac{y}{R}}$$

Q.29). Explain with short notes Prandtl mixing length theory

"Prandtl mixing length is that distance in the transvers direction which must be covered by a lump of fluid particles travelling with its original mean velocity in order to make the difference between its velocity and the velocity in new lamina EQUAL to the mean transverse fluctuation in turbulent flow."

Time – average of the absolute value of the fluctuation :

$$|\bar{u}| = \frac{1}{2}(|\Delta u_1| + |\Delta u_2|)$$

$$= \ell \left|\frac{du}{dy}\right|$$

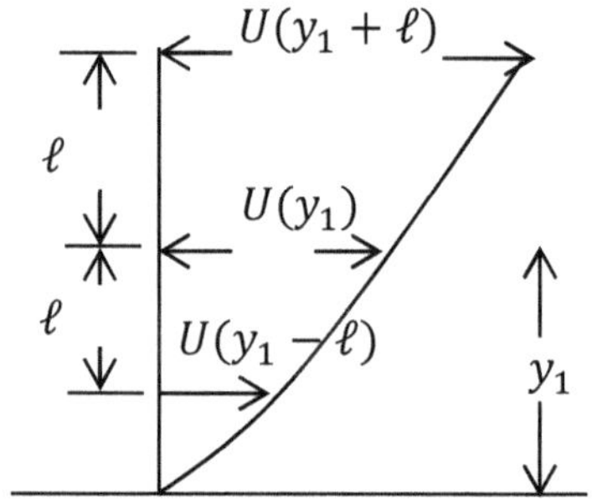

If the transverses components v′ is the same order of magnitude as u' and we put

$$|\bar{v'}| = Const. |\bar{u'}| = Const. \ell \left|\frac{dU}{dy}\right|$$

Then

$$|\overline{v'u'}| = - C|\bar{u'}|\,|\bar{v'}| \qquad \text{with } 0 < C < 1$$

$$\therefore \ |\overline{v'u'}| = - C\,\ell^2 \left(\frac{dU}{dy}\right)^2 \qquad \text{for } \ C = 1$$

$$|\overline{v'u'}| = \ - \ell^2 \left(\frac{dU}{dy}\right)^2$$

∴ shear stress can be written as

$$\therefore\ \tau_t = \rho\,\ell^2 \left(\frac{dU}{dy}\right)^2 \qquad \{\tau_t = -\,\rho\,\mathrm{v}'u'\}$$

$$\text{or}\ \ \tau_t = \rho\,\ell^2 \left|\frac{dU}{dy}\right| \left(\frac{dU}{dy}\right)$$

this is Prandtls' mixing length hypothesis comparing this equation with

$$\tau_t = \rho\,\varepsilon \left(\frac{dU}{dy}\right)$$

The expression for the eddy viscosity ε

$$\varepsilon = \ell^2 \left|\frac{dU}{dy}\right|$$

$$\text{or}\ \ \mu_t = \rho\,\ell^2 \left|\frac{dU}{dy}\right|$$

Q.30). Explain with short notes the effect of rough wall on laminar and turbulent flow:

Experimentally found that the surface roughness had no effect on friction resistance for Laminar flow. But turbulent flow strongly affected by roughness. The linear viscous sub layer only extends out to the nondimensionalized variable

$$y^+ = \frac{y\,u^*}{\upsilon} = 5 \qquad u^+ = \frac{u}{u^*}$$

Thus compared with diameter, the sublayer thickness y_s is only

$$\frac{y_s}{d} = \frac{5\,\upsilon/u^*}{d} = \frac{14.1}{Re_d\, f^{1/2}}$$

$$\text{For example, at } Re_d = 10^5, \qquad f = 0.018, \qquad \frac{y_s}{d} = 0.001$$

A wall Roughness of about 0.001 d will break up the sub – layer and profoundly change the wall law.

Prandtls' student Nikurade show, as in the following figure that the roughness height $\in$ will force the logarithm – law profile outward on the abscissa by an a mount approximately equal to $\in^{+}= \in u^{*}/v$. the slop of the logarithm law remains the same 1/k but the shift outward causes the constant B to be less by an amount of

$$\Delta B = \frac{1}{k} \ln \in^{+}$$

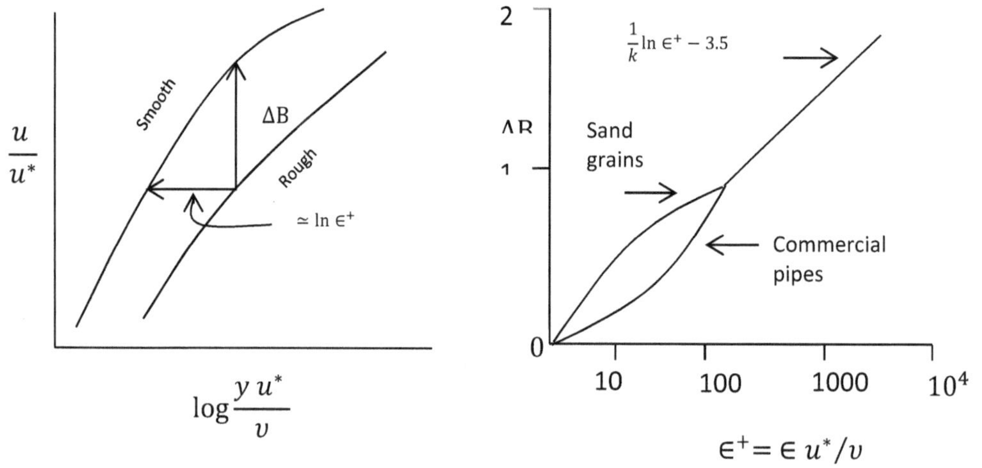

From Figure three regimes of rough walls

$\in u^{*}/v\ < 5$ hydraulically smooth walls, no effect of roughness on friction

$5 < \in u^{*}/v\ < 70$ transitional roughness, moderate Reynolds No. effect

$\in u^{*}/v\ < 70$ fully rough flow, sublayer totally broken up and friction independent of Reynolds No.

Q.31). Explain with short nots that diffusivity as far as application conserved helps to :

- Prevent B.L. separation
- Increase heat transfer rate
- Resist flow in pipelines
- Increase momentum transfer between wids and ocean current efc.

Q.32). Show that

1- For small scale turbulence the parameters governing the dissipation rate per unit mass $\in\ m^2/s^3$ and kinematic viscosity $\upsilon\ m^2/s$, can form length η, time Ţ and velocity V scale these scales are referred to as the Kolmogorov micro scales. State the formula of each?

2- Estimate the energy dissipation rate $\in$ for cumulus cloud per mass and for cloud height L = $1000\ m$ and velocity $u = 0.3\ m/s$.

3- Calculate the length, time and velocity scale for kinematic viscosity v = $15 \times 10^{-6}\ m^2/s$.

Solution:

1- $\eta = \left(\frac{\upsilon^3}{\in}\right)^{1/4}, \tau = \left(\frac{\upsilon}{\in}\right)^{1/2}, V = (\upsilon \in)^{1/4}$

2- $\in = \frac{u^3}{\ell} = \frac{(0.3)^3}{1000} = 2.7 \times 10^{-5}\ m^2/s^3$

3- $\eta = \left(\frac{(15\times10^{-6})^3}{2.7\times10^{-5}}\right)^{1/4} = 3.34 \times 10^{-3}$

$$\tau = \left(\frac{15\times10^{-6}}{2.7\times10^{-5}}\right)^{1/4} = 0.745$$

$$V = (15 \times 10^{-6} \times 2.7 \times 10^{-5})^{1/4} = 4.5 \times 10^{-3}$$

Q.33). The total stress state : $\text{Ţ} = \mu \frac{\partial u^-}{\partial y} - \rho\, u'v'$, Explain the relation of each part with the pressure gradient and velocity gradient at the wall.

Solution:

$$\tau = \mu \frac{\partial u}{\partial y} - \rho\, u'v'$$

Where

$\mu \dfrac{\partial u}{\partial y}$ is the viscous shear

and

$-\rho\, u'v'$ is the Reynolds stress

Experimental results provide important guide lines for the anaylysis of the equation. The relative magnitudes of the viscous and Reynolds stresses across a B.L. can be abtained experimentally. The following sketches show the resulting typical stress distributions for (1) zero pressure gradient (2) favourable pressure gradient and (3) adverse pressure gradient.

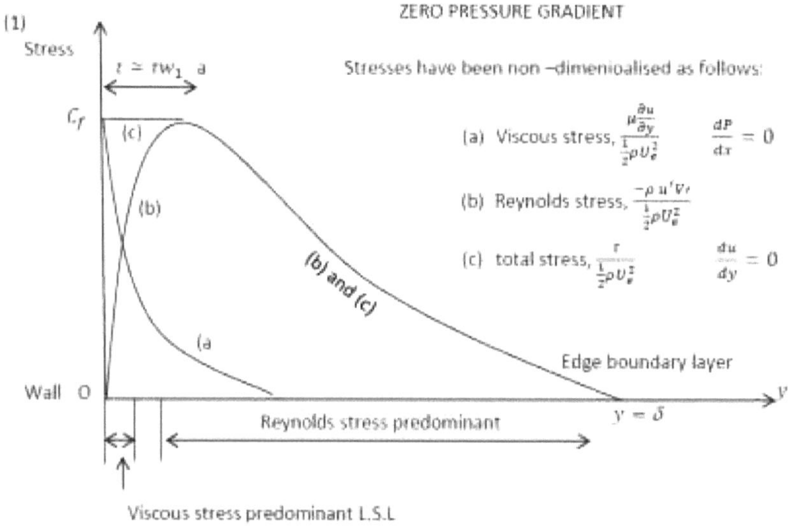

(2)

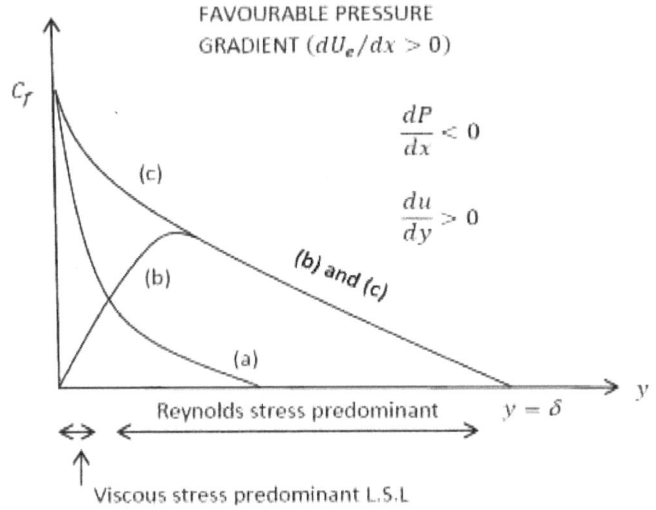
FAVOURABLE PRESSURE
GRADIENT $(dU_e/dx > 0)$
C_f
$\frac{dP}{dx} < 0$
$\frac{du}{dy} > 0$
(c)
(b)
(b) and (c)
(a)
y
Reynolds stress predominant
$y = \delta$
Viscous stress predominant L.S.L

(3)

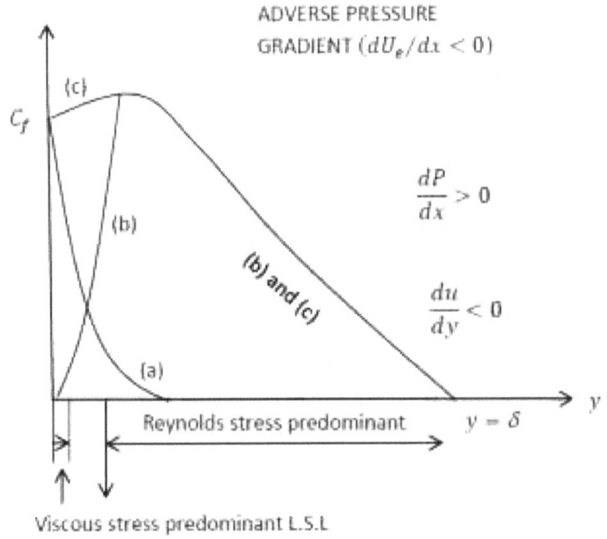
ADVERSE PRESSURE
GRADIENT $(dU_e/dx < 0)$
C_f
(c)
(b)
(b) and (c)
(a)
$\frac{dP}{dx} > 0$
$\frac{du}{dy} < 0$
y
Reynolds stress predominant
$y = \delta$
Viscous stress predominant L.S.L

In all three case there is a thin layer adjucent to the wall in which the viscous stress in much larger than the Reynolds stress. This is because $u' \to 0$ and $v' \to 0$ as $y \to 0$ by the no – slip conditions and so the Reynolds stress, $-\rho\, u'v' \to 0$ at $y \to 0$. This thin region is called the laminer sublayer. The sketches shows that Reynolds stress are predomenat over the majer parts of the B.L. it is also apperent that $\tau = \tau_w$ for smaal pressure gradients and small values of y.

Q.34). State the Causes of transients flow?

1- Opening, closing or " chattering " of valves in a pipeline.
2- Starting or stopping the pums in apumping system.
3- Starting up a hydraulic turbine, accepting or rejecting loud.
4- Vibrations of the vanes of a runner or an impeller, or of the blades of a fan.
5- Sudden changes in the inflow or outflow a canal by opening or closing the control gate .
6- Faiulure or collapse of a dam.
7- Sudden incrsases in the inflow to a river on a sewer due to flush storm runoff.

Q.35). Drive the continuity equation due to transient's flow

Answer

Consider the C.V. shown below.

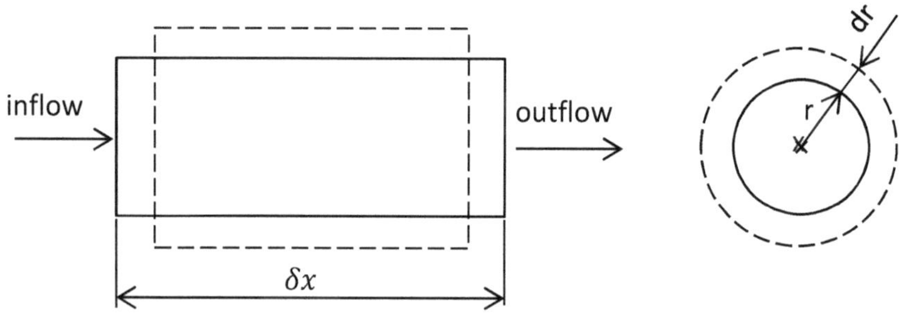

The volume of fluid inflow V_{in} and outflow V_{out} during time interval δt are

$$V_{in} = Q\delta t = V\pi r^2 \delta t \quad \textbf{(1)}$$

and $V_{out} = \left(V + V\frac{\partial V}{\partial x}\delta x\right)Tr^2\delta t$ **(2)**

where r = radius of the conduit.

The increase in the fluid volume during δt

$\therefore \ \delta V_{in} = V_{in} - V_{out}$

$= -\frac{\partial V}{\partial x}\delta x \delta t \pi r^2$ **(3)**

the pressure change, δP during time interval δt is $\left(\frac{\partial P}{\partial t}\right)\delta t$.

This pressure change causes the conduit wall to expand or contract radially and causes the length of the fluid element to decrease or increase due to fluid compressibility.

Let us first consider the volume change δV_r, due to the radial expansion or contraction of the conduite.

The radial or hoop stress, σ, in a conduit due to the pressure P is given by the following:

If one – help of this ring is taken as afree body

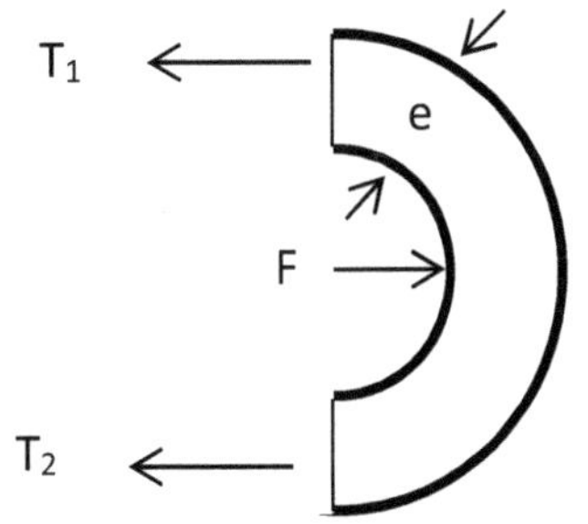

$\delta r = e$ for wall thickness

The horizontal componets of force acts through the pressure center of the projected area

$$F = PA = p2r \times 1$$

Since $$T_1 = T_2$$

$\therefore$ $$2T = 2rp$$

$\therefore$ $$T = pr \qquad \textbf{(4)}$$

In which T is thensile force per unit length.

For wall thickness e, the tensile stress in the pipe wall is

$$\sigma = \frac{T}{e} = \frac{pr}{e} \qquad \textbf{(5)}$$

Hence, the change in hoop stress, $\delta\sigma$ causd by δP may be written as

$$\delta\sigma = \frac{\partial P}{\partial t} . \delta t \frac{r}{e} \qquad \textbf{(6)}$$

and the change in strain $\delta \in = \frac{\delta\sigma}{r}$ **(7)**

the canduit walls are assumed linearly elastic, then

$E = \frac{\delta\sigma}{\delta\in}$ youngs modulus of elasticity **(8)**

Subsituting eqs. (6) and (7) in (8) we get

$$E = \frac{(\partial P/\partial t)\delta t\left(\frac{r}{e}\right)}{\delta r/r} \quad \textbf{(9)}$$

Or

$$\delta r = \frac{\partial P}{\partial t} \cdot \frac{r^2}{eE}\ \delta t \quad \textbf{(10)}$$

The change in volume in radial expansion

$$\delta V_r = 2\pi r \delta x \delta r \quad \textbf{(11)}$$

Or

$$\delta V_r = 2\pi \frac{\partial P}{\partial t} \cdot \frac{r^3}{eE}\ \delta t\ \delta x \quad \textbf{(12)}$$

The change in volume due to compressibility of fluid δV_c and the initial volume of fluid

$$V = \pi r^2\ \delta x \quad \textbf{(13)}$$

The bulk modulus of elasticty of fluid K is definds' by

$$K = \frac{-\ \delta P}{\frac{\delta V_C}{V}} \quad \textbf{(14)}$$

From (13) in (14) and $\delta P = \dfrac{\partial P}{\partial t}\ \delta t$

$$\therefore \; \delta V_c = \frac{-\partial P}{\partial t} . \frac{\delta t}{K} \pi r^2 \delta x \quad \textbf{(15)}$$

If we assume fluid density is constant then

$$\delta V_{in} + \delta V_c = \delta V_r \quad \textbf{(16)}$$

Substituting eqs. (3,12,15) in to eq. (16) and dividing both side by $\pi r^2 \delta x \delta t$ yield

$$\frac{-\partial V}{\partial x} \delta x \delta t \pi r^2 + \frac{-\partial P}{\partial t} . \frac{\delta t}{K} \pi r^2 \delta x = 2\pi \frac{\partial P}{\partial t} . \frac{r^3}{eE} \delta t \, \delta x$$

$$-\frac{\partial V}{\partial x} - \frac{\partial P}{\partial t} . \frac{1}{K} = 2 \frac{\partial P}{\partial t} . \frac{r}{eE}$$

or $\frac{\partial V}{\partial x} + \frac{\partial P}{\partial t}\left[\frac{1}{K} + \frac{2r}{eE}\right] = 0$ **(17)**

Since

$Q = AV$ then $\mathrm{V} = \frac{Q}{A}$ and $P = \gamma H$

$\frac{\partial V}{\partial x} = \frac{\partial Q}{\partial x} . \frac{1}{A}$ and $\frac{\rho g \partial H}{\partial t} = \frac{\partial P}{\partial t}$

$\frac{1}{A}\frac{\partial Q}{\partial x} + \frac{\rho g \partial H}{\partial t}\left[\frac{1}{K} + \frac{D}{eE}\right] = 0$ multiply both side by $\frac{K}{K}$

$\frac{1}{A} . \frac{\partial Q}{\partial x} + \rho g \frac{\partial H}{\partial t}\left[1 + \frac{KD}{eE}\right] \times \frac{1}{K} = 0$ **(18)**

Multiply both side by $\left[\frac{K}{1+\frac{KD}{eE}}\right]\frac{1}{\rho g}$

And let us define $a^2 = \frac{K}{\rho\left[1+\frac{KD}{eE}\right]}$

Then equation (18) becomes

$$\frac{a^2}{gA}\frac{\partial Q}{\partial x} + \frac{\partial H}{\partial t} = 0 \quad \text{continuity eq.}$$

Where a is the wave velocity

Q.36). Explain the rate of change of disturbance energy of fluid particles during trasition zone.

1- The rate of work done by the additional stress on the boundery surface of the particale due to disturbance velocities (say.A). this may either increase or decrease the disturbance energy.

2- The rate at which the energy of the disturbing motion is dissipated by viscosity (say.B). this will tend to decrcase the disturbance energy of the partical.
If Re is sufficiently small, then B > A and so the disturbance are damped out by viscosity, resulting in stable flow

Q.37). Explain the effect of wave number α and wave length ℓ on the disturbance and stability of flow ? where $\bar{\alpha} = L\alpha$.

1- If Re < Re_{crit} the flow is stable for all volues of $\bar{\alpha}$ (*i.e.* all disturbance). The viscous force dump out all disturbance.

2- If the wave number α is large (*i.e.* small wave length ($\ell = \frac{2\pi}{\alpha}$)), the flow is stable for all Re. the disturbance velocity have large gradients (*e.g*) $\frac{\partial u'}{\partial y}$ large causing the disturbance to be damping by viscosity.

3- If U has a point of inflexion, there exists a wave number α_0 say such that the flow is unstable if $\alpha < \alpha_0$ and $R \to \infty$

The stability curves depends only on the lacal velocity profile and so do not take account of upstream influence as shown in the following figure.

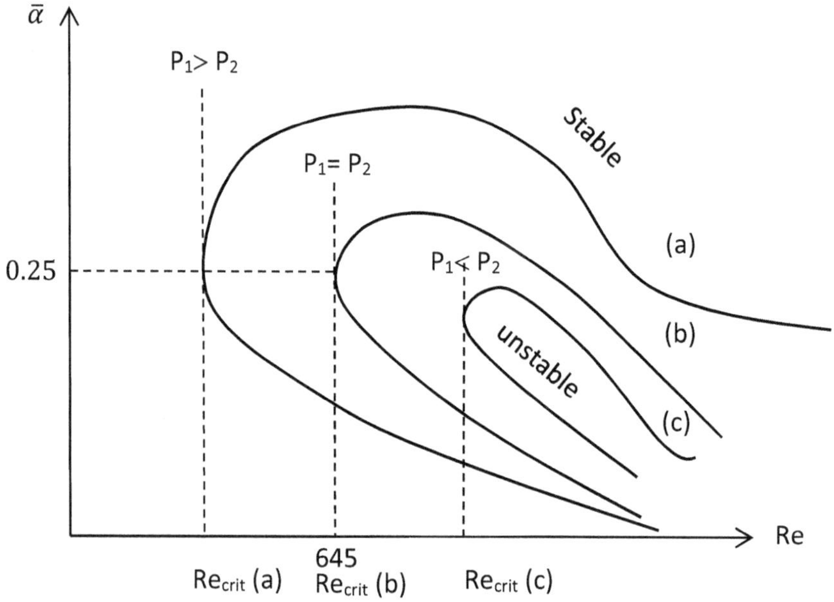

Curve of neutral stabilty for

a) Adverse pressure gradient and point of inflexion in the velocity profile.

b) Zero pressure gradient.

c) Favoroble pressure gradient (Change the pressure gradient to get the stability).

Q.38). Estimate the value of the eddy kinematic viscosity ∈ as a function of position for turbulent flow of water in a smooth pipe of internal diameter 100 mm. The centerline velocity is 6.1 m/s and the pressure droop over a length of 8 m is 14.000 Pa. For water, the value of the kinematic viscosity υ is $1 \times 10^{-6}\ m^2/s$. It may be assumed that the mean velocity profile is described by Prandtl 1/7 the power law over most of the cross – section.

Solution:

The eddy kinematic ∈ is defined by

$$\bar{\tau}_{rx} = -\rho(\upsilon + \in)\frac{\partial v_x}{\partial r}$$

Prandtl 1/7 power law is

$$\frac{\upsilon_x}{V'_{max}} = \left(\frac{y}{r_i}\right)^{\frac{1}{7}}$$

Where y is the distance measured from the wall.

Thus

$$V'_x = V_{max}\left(1 - \frac{r}{r_i}\right)^{\frac{1}{7}}$$

And

$$\frac{dV_x}{dr} = \frac{V_{max}}{7\,r_i}\left(1 - \frac{r}{r_i}\right)^{\frac{1}{7}} = -17.43\left(1 - \frac{r}{r_i}\right)^{-\frac{6}{7}} \delta^{-1}$$

The shear stress at the wall given by

$$\tau_w = \frac{d_i}{4}\left(\frac{\Delta P_i}{L}\right) = \frac{0.1 \times 1400}{4 \times 8} = 43.75\ Pa$$

At radial distance r, the shear stress is given by

$$\frac{\bar{\tau}_{rx}}{\tau_w} = \frac{r}{r_i}$$

Using these equations, can be constructed

r/r_i	$\tau(Pa)$	dV_x/dr	$\in m^2/s$	$\in/\upsilon$
0.1	4.38	-19.1	0.23×10^{-3}	230
0.2	8.75	-21.1	0.41×10^{-3}	410
0.4	17.5	-27.0	0.63×10^{-3}	650
0.6	26.25	-38.2	0.68×10^{-3}	680
0.8	35.0	-69.3	0.51×10^{-3}	510
0.9	39.38	-125	0.31×10^{-3}	310

These values of $\in$ may be compared with the value of the molecular kinematic viscosity $\upsilon(1 \times 10^{-6}\ m^2/s)$ $\in$ is nearly there orders of magnitude larger than υ. At the wall $\in \rightarrow 0$, but this behavior cannot be calculated from 1/7 th power law.

Part II

Turbulent Boundary Layer

CHAPTER 5

Boundary Layer

Abstract: Several exact solutions considered in literature notably the moving – boundary flows, stagnation flow, the rotating disk, convergent – wedge flow, and the flat plate with asymptotic suction – have hinted strongly at boundary – layer behavior. That is, at large Reynolds numbers the effect of viscosity becomes increasingly confined to narrow regions near solid walls. The computer models solutions also shewed this tendency at large Re to sweep the vorticity downstream and leave the flow far from the walls essentially irrational. Physically, this means that the rate of downstream convections is much larger than the rate of transverses viscous diffusion.

In this chapter, definitions related to the Boundary layer are described by a formula. A brief description and derivation in details of Boundary layer.

Theories, integral momentum equation of boundary layer implemented on flat – plate for laminar and turbulent flow, simulation solution for steady 2D flow, Blasius solution, Falkner – Skan wedge flow, and Thwaite's method.

Keywords: Boundary layer, Blasius solution, Falkner–Skan wedge flow, Flat plate, Thwaite's method.

5.1. INTRODUCTION

Boundary – Layer theory is cornerstone of our knowledge of flow of air and other fluids of small viscosity under circumstances of interest in many engineering applications.

The flow of real fluid (except at extremely low pressure) has two fundamental characteristics one is that there is no discontinuity of velocity second is that at a solid surface, the velocity of the fluid relative to the surface is zero. As a result, there is close to the surface a region in which the velocity increases rapidly from zero and approaches the velocity of the boundary layer.

The B.L. which is simplest to study is that formed in the flow along one side of a thin, smooth flat plate parallel to the direction of the on comming fluid as shown in Fig. (**5.1**). [1, 7].

Jafar Mehdi Hassan, Riyadh S. Al-Turaihi, Salman Hussien Omran, Laith Jaafer Habeeb, Alamaslamani Ammar Fadhil Shnawa

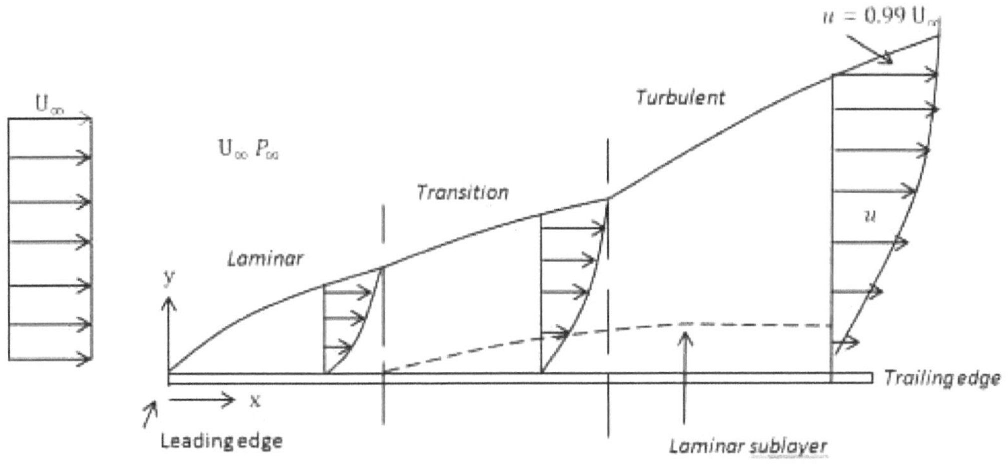

Fig. (5.1). Boundary Layer growth along flat plate [1].

No other solid surface is near, and the pressure of the fluid is assumed uniform. The velocity gradient in a real fluid are therefore entirely due to viscose action near the surface.

The thickness of the layer may be taken as that distance from the surface at which the velocity reaches 99% of the velocity of the main stream.

The flow in the first part is entirely laminar, whether or not the main flow is laminar with increasing thickness, however the laminar layer becomes unstable and the motion within it become disturbed. These change take place over a short length known as the transition zone. Downstream of the transition region the B.L. is almost entirely turbulent and its thickness increase further.

At any distance x from the leading edge of the plate the boundary thickness δ is very small compared with x.

In a turbulent layer there is more intermingling of fluid particles and therefore a more nearly uniform velocity than in a laminar layer as in Fig. (**5.2**).

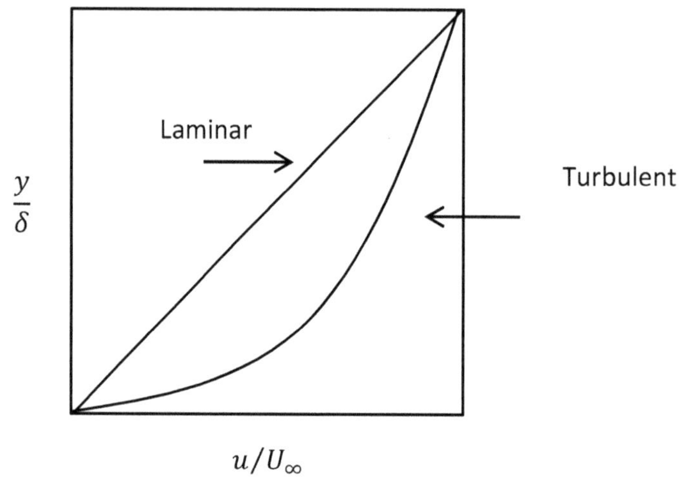

Fig. (5.2). Velocity determination.

The point at which a laminar B.L. becomes unstable depends on a number of factors.

1. Roughness of surface
2. Reynolds number. $Re_x = \frac{x.V}{\upsilon}$ where 5.0×10^5 up to 5.45×10^5 the transition zone.

When P is not uniform

$$\frac{dP}{dx} > 0, \quad Re_c \rightarrow \text{ transition occurs lower}$$

$$\frac{dP}{dx} < 0, \quad Re_c \rightarrow \text{ transition occurs higher}$$

In general, B.L. be laminar over only a relative short distance from the leading edgy and then it is often assumed with sufficient accuracy that the layer as turbulent over its entire length.

5.2. DEFINITIONS

B.L. thickness: The velocity within the B.L. increases to the velocity of the main stream asymptotically, some arbitrary conversion must be adopted to define the thickness of the layer.

One of these is the displacement thickness.

5.2.1. Displacement Thickness

The area under the velocity profile, Fig. (**5.3**) (under the curve) is

$$Area = \int_{0}^{\infty} (U_{\infty} - u)\, dy$$

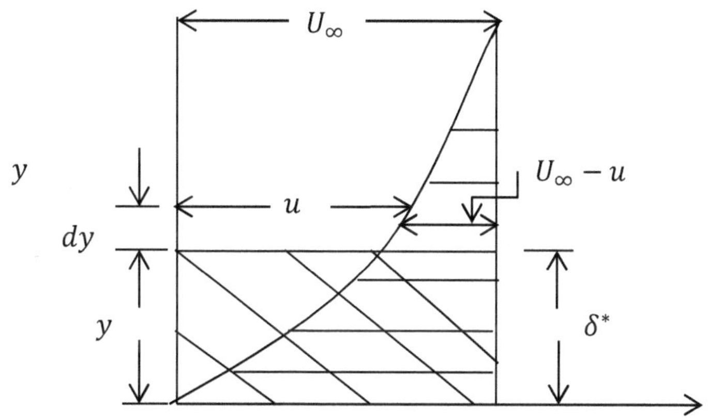

Fig. (5.3). Displacement thickness.

a- For real fluid, the volume flow rate per unit depth $= u\, dy$

b- For ideal fluid, (inviscid flow $\mu = 0$) the volume flow rate per unit depth

$$= U_{\infty}\, dy$$

$$\therefore\ \Delta Q = Q_2 - Q_1$$

$$= U_\infty\, dy - u\, dy = U_\infty \delta^*$$

$$\text{or } \delta^* = \frac{1}{U_\infty}\int_0^\infty (U_\infty - u)\, dy$$

$$\therefore\ \delta^* = \int_0^\infty \left(1 - \frac{u}{U_\infty}\right) dy \qquad (5.1)$$

5.2.2. Momentum Thickness θ

By the same manner

a- For real fluid $= \rho u\, dy. u$

b- For ideal fluid $= \rho u\, dy. U_\infty$

or $\rho u\,(U_\infty - u)dy = \rho U_\infty \theta. U_\infty$

$$\therefore\ \theta = \int_0^\infty \frac{u}{U_\infty}\left(1 - \frac{u}{U_\infty}\right) dy \qquad (5.2)$$

5.2.3. Kinetic Energy Factor

Also

$$\delta^{**} = \int_0^\infty \left(\frac{u}{U_\infty}\right)^2 \left(1 - \left(\frac{u}{U_\infty}\right)^2\right) dy \qquad (5.3)$$

5.2.4. Shape Factor

$$H = \frac{\delta^*}{\theta} \qquad (5.4)$$

5.3. THE SEPARATION OF A B.L.

If we apply the momentum equation at the wall where

$u = \mathrm{v} = 0$ we find that

$$\left.\frac{\partial^2 u}{\partial y^2}\right\}_{y=0} = \frac{1}{\mu}\frac{dP}{dx} \tag{5.5}$$

That the wall curvature has the sign of the pressure gradient, whereas further out the profile must have negative curvature when it merges with the free stream. Profile curvature is an indicator of possible B.L. separation. There examples are shown in the following Fig. (**5.4**). For negative (favorable) pressure gradient, the curvature is negative throughout and no flow separation can occur for zero gradient *e.g.* flat – plate flow, the curvature is zero at the wall and negative further, out there is no separation. For positive (adverse) gradient, the curvature changes sign and the profile in s- shaped. The increasing downstream pressure slows down the wall flow and can make it go backward – flow separation.

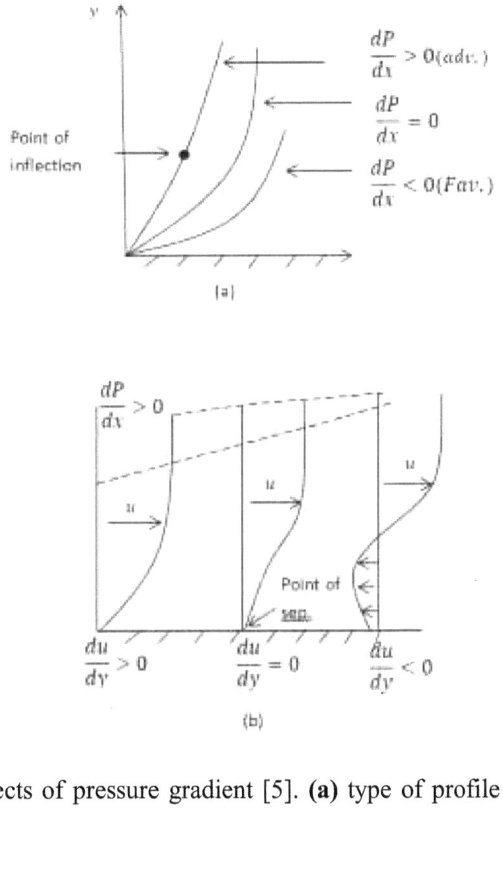

Fig. (5.4). Geometric effects of pressure gradient [5]. **(a)** type of profile **(b)** persistence adverse gradient.

Fig. (**5.4b**) illustrates the separation process. A persistent adverse gradient $(dP/dx > 0)$ makes the profile more and more s-shaped, reducing the wall shear to zero (the separation point) and then causing backflow, while the B.L. becomes much thicker, laminar flows have poor resistance to adverse gradients and separate easily. Turbulent B.L. can resist separation longer at the expense of increased wall friction and heat transfer.

Also $\frac{\partial^2 u}{\partial y^2} = 0$ at a large distance from the wall

5.4. PRESSURE DRAG

When flow occurs past a surface which is not everywhere parallel to the main stream, there is an additional drag force resulting from difference of pressure over the surface. This force known as pressure drag. Which depends on the form of B.L. [6].

The skin friction drag is the resultant of the tangential forces, while the pressure drag is a resultant of a normal forces.

The sum of the skin and pressure drag is termed the profile drag.
Profile drag = skin friction drag + pressure drag.

Downstream of the separation position the flow is greatly disturbed by large – scale eddies and this regain of eddying motion is usually known as the wake.
As a result of the energy disputed by the highly turbulent motion in the wake, the pressure reduced and the pressure drag increased.

Pressure drag depends on the size of the wake and position of separation (depends on the body shape) as shown in Fig. (**5.5**).

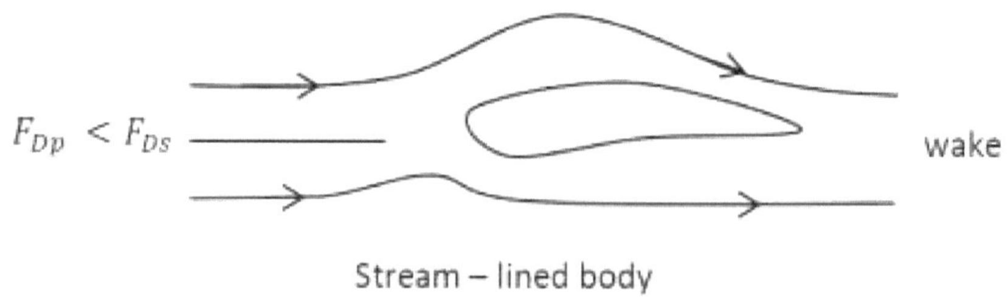

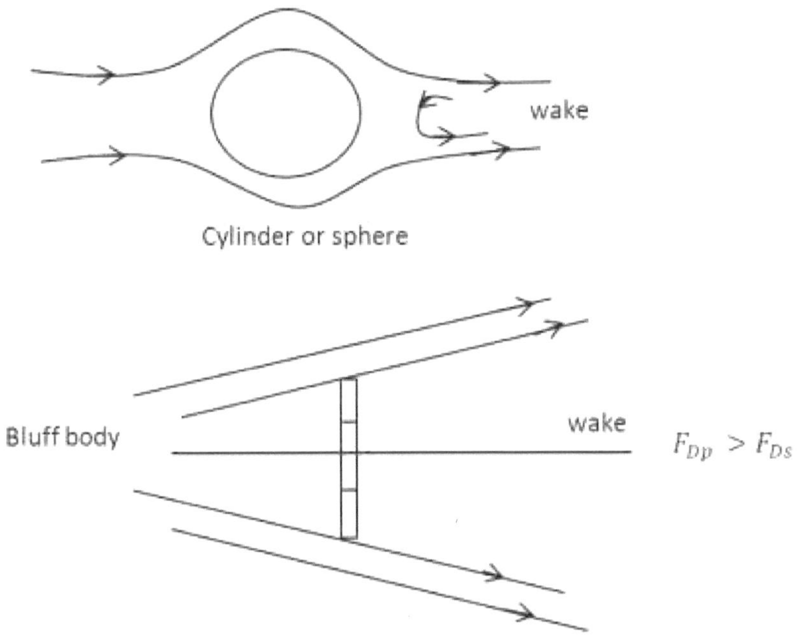

Fig. (5.5). Flow past different types of bodies illustrating the pressure drag and the wake zone [5].

5.5. BOUNDARY LAYER THEORIES

5.5.1. Integral Momentum Equation of B.L.

To study the momentum equation for steady flow in a B.L. on a flat plate over which there may be a variation of pressure in the direction of flow, we derived the momentum integral relation for a fixed control volume over a flat plate boundary layer [8]. In Fig. (**5.6**) a control volume is taken enclosing the fluid above the plate extending the distance x along the plate.
$AE = \delta x$ of the plate. The width is assumed large so that edge effects are negligible and the flow 2D.

The B.L. thickness δ and its outer edge is represented by (BD).

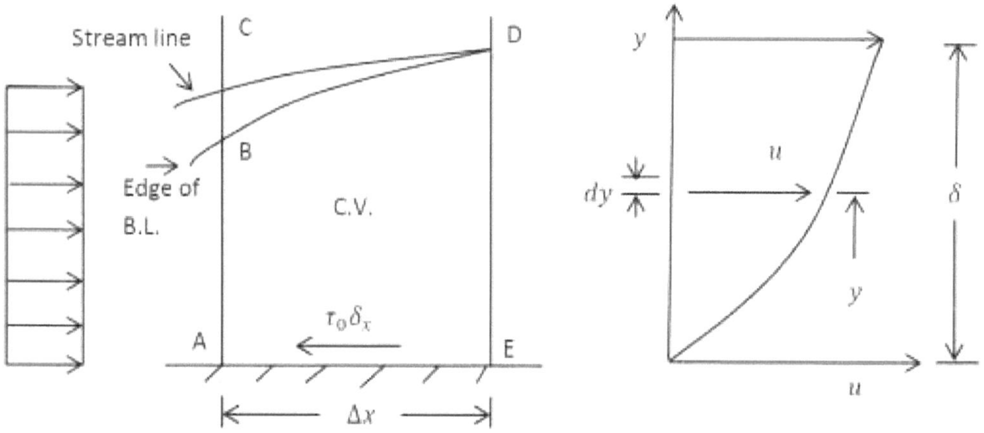

Fig. (5.6). Control volume applied to fluid flow over one side of a flat plate.

Let C.be the point on AB produced which is on the same streamline as D.

No fluid crosses the streamline CD. Then ACDE is a control volume.

Then $\sum$F = change in momentum.

$$\sum F = momentum\ out - momentum\ in\ C.V. \quad \textbf{(5.6)}$$

1- left hand side of the above equation, let the pressure over the face AC to have the main P

at the face ED the pressure is $\left(P + \frac{\partial P}{\partial x} dx\right)$

For unit width the difference of pressure therefore produce on the $C.V. = P(AC) - \left(P + \frac{\partial P}{\partial x} dx\right) ED + \left(P + \frac{1}{2}\frac{\partial P}{\partial x} dx\right)(ED - AC)$ **(5.7)**

where $\left(P + \frac{1}{2}\frac{\partial P}{\partial x} dx\right)$ is the mean value of pressure over the surface CD

Equation (5.6) reduce to $-\frac{1}{2}\frac{\partial P}{\partial x} dx\ (ED - AC)$

When $dx \rightarrow 0$ and $AC \rightarrow ED$ in magnitude then the expression above becomes $= \left(-\frac{\partial P}{\partial x} dx\right) ED$ and $ED \rightarrow \delta$

Then the total force in x – dir.

$$= -\tau_0 dx - \frac{\partial P}{\partial x} dx(ED) \quad \textbf{(5.8)}$$

Where τ_0 represent the shear term.

2- The right-hand side of eq. (5.7), through the elementary strip in the plan, AB, distance AB, distance y from the surface.

Surface	**Momentum ($\dot{m}$v)**	**Direction**
AB	$\int_0^{\delta} \rho u^2\, dy$	in flow
BC	$\rho UU(BC)$	in flow
ED	$\int_0^{\delta} \rho u^2\, dy + \frac{\partial}{\partial x}\left(\int_0^{\delta} \rho u^2\, dy\right)\delta x$	out flow

mass flow across BC = mass flow across ED – mass flow across AB

$$\rho U(BC) = \int_0^{\delta} \rho u\, dy + \frac{\partial}{\partial x}\left(\int_0^{\delta} \rho u\, dy\right) dx - \int_0^{\delta} \rho u\, dy$$

$$= \frac{\partial}{\partial x}\left(\int_0^{\delta} \rho u\, dy\right) dx \quad \textbf{(5.9)}$$

From Table above

$$\Sigma Fx = \int_0^\delta \rho u^2\, dy + \frac{\partial}{\partial x}\left(\int_0^\delta \rho u^2\, dy\right) dx - \int_0^\delta \rho u^2\, dy - \rho UU(BC) \qquad \textbf{(5.10)}$$

Substituting eq. (5.9) in (5.10) for BC and rearrange eq. (5.10) we get

$$\Sigma Fx = \frac{\partial}{\partial x}\left(\int_0^\delta \rho u^2\, dy\right) dx - U\frac{\partial}{\partial x}\left(\int_0^\delta \rho u\, dy\right) dx \qquad \textbf{(5.11)}$$

From eqs (5.8) and (5.11) we get

$$-\tau_0 dx - \frac{\partial P}{\partial x} dx \int_0^\delta dy = \frac{\partial}{\partial x}\left(\int_0^\delta \rho u^2\, dy\right) dx - U\frac{\partial}{\partial x}\left(\int_0^\delta \rho u\, dy\right) dx \qquad \textbf{(5.12)}$$

From Bernoulli eq.

$$P + \frac{1}{2}\rho U^2 = c \qquad \text{then} \qquad \frac{\partial P}{\partial x} + \rho U\frac{\partial U}{\partial x} = 0$$

$$\text{or} \quad \frac{\partial P}{\partial x} = -\rho U\frac{\partial U}{\partial x} \qquad \textbf{(5.13)}$$

Substituting eq. (5.12) → (5.11) we get

$$-\tau_0 + \rho U\frac{\partial U}{\partial x}\int_0^\delta dy = \frac{\partial}{\partial x}\left(\int_0^\delta \rho u^2\, dy\right) - U\frac{\partial}{\partial x}\left(\int_0^\delta \rho u\, dy\right) \qquad \textbf{(5.14)}$$

Where

$$U\frac{\partial}{\partial x}\int_0^\delta \rho u\, dy = U\frac{\partial u}{\partial x}\int_0^\delta \rho\, dy$$

$$= \left(\frac{\partial}{\partial x}Uu - u\frac{\partial U}{\partial x}\right)\int_0^\delta \rho\, dy$$

$$= \rho\frac{\partial}{\partial x}\int_0^\delta Uu\, dy - \rho\frac{\partial U}{\partial x}\int_0^\delta u\, dy \qquad \textbf{(5.15)}$$

Substituting eq. (5.15) in eq. (5.14) we get

$$\tau_0 = -\frac{\partial}{\partial x}\left(\int_0^\delta \rho u^2\, dy\right) + \rho\frac{\partial}{\partial x}\int_0^\delta Uu\, dy - \rho\frac{\partial U}{\partial x}\int_0^\delta u\, dy + \rho U\frac{\partial U}{\partial x}\int_0^\delta dy$$

$$\tau_0 = \rho\frac{\partial}{\partial x}\left(\int_0^\delta Uu\, dy - \int_0^\delta u^2\, dy\right) + \rho\frac{\partial U}{\partial x}\left(\int_0^\delta U\, dy - \int_0^\delta u\, dy\right)$$

$$\therefore\ \tau_0 = \rho\frac{\partial}{\partial x}\int_0^\delta (U-u)\, u dy + \rho\frac{\partial U}{\partial x}\int_0^\delta (U-u) dy \quad \textbf{(5.16)}$$

This equation is a general equation used in all the zone of B.L. (Laminar, Turbulent) for open zone and flat plate the effect of pressure variation could be neglecting.

5.5.2. B.L. on Flat Plate

Assuming 1. Steady flow 2. 2 – dim 3. $\frac{\partial P}{\partial x} = 0$ 4. $\frac{\partial U}{\partial x} = 0$
Then eq. (5.16) becomes:

$$\tau_0 = \rho\frac{\partial}{\partial x}\int_0^\delta (U-u)\, u dy$$

or $\tau_0 = \rho U^2 \frac{\partial}{\partial x}\int_0^\delta \frac{u}{U}\left(1-\frac{u}{U}\right) dy$ **(5.17)**

at $y = 0 \quad u = 0$
$y = \delta \quad u = U$

For dimensionless solution, the velocity distribution

$\frac{u}{U} = f\left(\frac{y}{\delta}\right) = f(\eta)$ and $\eta = \frac{y}{\delta}$

let B.C. $\eta = 0 \rightarrow f = 0 \quad u = 0 \quad y = 0$

$\eta = 1 \rightarrow f = 1 \quad u = U \quad y = \delta$

or $\delta\eta = y$ and $\delta d\eta = dy$

then eq. (5.16) becomes

$$\tau_0 = \rho U^2 \frac{\partial \delta}{\partial x} \int_0^1 f(\eta)(1 - f(\eta))\, d\eta \quad \textbf{(5.18)}$$

let $\int_0^1 f(\eta)(1 - f(\eta))\, d\eta = A$

$$\therefore\ \tau_0 = \rho U^2 A \frac{\partial \delta}{\partial x} \quad \textbf{(5.19)}$$

5.5.2.1. Laminar B.L.

In practical the laminar B.L. part is often so short that it may be neglected especially at low values of Reynolds No.

For laminar flow the shear stress at the wall

$$\tau_0 = \mu \left.\frac{du}{dy}\right)_{y=0} \quad \textbf{(5.20)}$$

Since $du = U df(\eta)$ $dy = \delta d\eta$

Then $\tau_0 = \frac{\mu}{\delta} U \left(\frac{\partial f(\eta)}{\partial \eta}\right)_{\eta=0}$

$$\tau_0 = \frac{\mu}{\delta} UB \quad \textbf{(5.21)}$$

When $B = \frac{\partial f(\eta)}{\partial \eta}$

From eq. (5.19) and (5.20) we get

$$\delta = \sqrt{\frac{2\mu Bx}{\rho UA}} \qquad \text{or} \qquad \delta = \sqrt{\frac{2B}{A}} \cdot \frac{x}{\sqrt{Re_x}} \tag{5.22}$$

$$\text{Also } \tau_0 = \rho U^2 \sqrt{\frac{AB}{2Re_x}} \tag{5.23}$$

$$F_D = \sqrt{2AB\mu\rho U^3 L} \tag{5.24}$$

When L: The length of the plate

$$\text{and } C_D = 2\sqrt{\frac{2AB}{Re_x}}$$

Ex. For Prandtl law for laminar B.L.

$$f(\eta) = \frac{3}{2}\eta - \frac{1}{2}\eta^3 \quad \text{lead to}$$

$$\frac{\delta}{x} = \frac{4.65}{\sqrt{Re_x}} \qquad \tau_0 = 0.322\sqrt{\frac{\mu\rho U^3}{x}}$$

$$F_D = 0.644\sqrt{\mu\rho U^3 L} \quad \text{per unit width}$$

$$C_D = \frac{1.288}{\sqrt{Re_x}} \tag{5.25}$$

5.5.2.2. Turbulent B.L.

Prandtl has suggested that

$$\frac{u}{U} = \left(\frac{y}{\delta}\right)^{\frac{1}{n}} = \eta^{\frac{1}{n}} \tag{5.26}$$

or $f(\eta) = \eta^{\frac{1}{n}}$

since eq. $\tau_0 = \mu \left.\frac{du}{dy}\right)_{y=0}$ is equal to (∞) for $f(\eta) = \eta^{\frac{1}{n}}$

then using eq. of $\tau = \frac{1}{2} C_f \rho V^2$

and for Blasius formula friction factor in pipes

$$f = 0.316\, Re_d^{-\frac{1}{4}} \quad \text{where} \quad Re_d = \frac{\rho D V}{\mu}$$

$U = 1.235$ V and $D = 2R = 2\delta$

From all $\tau_0 = 0.0228\, \rho U^2 Re_\delta^{-\frac{1}{4}}$ **(5.27)**

δ, τ_0, F_D, C_D ⋯ for any range of Reynolds No.

5.5.3. Simulation Solution for Steady 2D. Flow

The simplest example of the application of the B.L. equ's is afforded by a flow along every thin flat plate [5], (Fig. **5.7**).

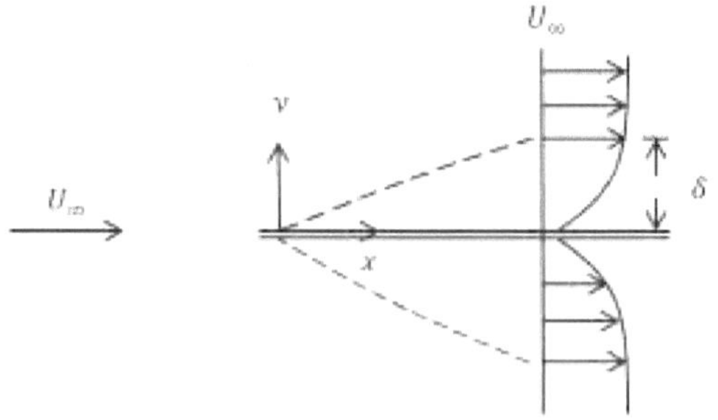

Fig. (5.7). Boundary layers on thin plate.

For steady flow with a free stream velocity U_∞. The continuity equation and momentum equations because:

$$\frac{\partial u}{\partial x} + \frac{\partial v}{\partial y} = 0 \qquad \textbf{(5.28)}$$

And

$$u\frac{\partial u}{\partial x}+\mathrm{v}\frac{\partial \mathrm{u}}{\partial y}=U\frac{\partial U}{\partial x}+\upsilon\frac{\partial^2 \mathrm{u}}{\partial y^2} \qquad \textbf{(5.29)}$$

Where the free stream outside the B.L., $U = U_{(x)}$ where x is the coordinate parallel to the wall, is related to $P_{(x)}$ by Bernoulli's theorem the incompressible flow.

$$\frac{\partial P}{\partial x}=-\rho U\frac{\partial U}{\partial x} \qquad \textbf{(5.30)}$$

Then $U_{(x,0)} = V_{(x,0)} = 0$ $\qquad u_{(x,\infty)} = U_{(x)}$

5.5.4. The BLASIUS Solution for Flat – Plate Flow

If the displacement thickness is small losses is very small, $(Re \gg 1)$ then $U =$ constant and $\frac{\partial U}{\partial x} = 0$ in eq.(5.29). Therefore, Blasius introduced a similarity variable η that combines independent variables, x and y into one nondimensionalized in depend variable and solved for a nondimensionalized form of the x – component of velocity and local velocity profile [1,5].

$\frac{u}{U} = f\left(\frac{y}{\delta}\right)$ and $\eta = \frac{y}{\delta}$

Since from integral analysis $\delta = const.\left(\frac{\upsilon x}{U}\right)^{\frac{1}{2}}$

The appropriate dimensionless similarity variable should be

$\frac{y}{\eta} = const.\left(\frac{\upsilon x}{U}\right)^{\frac{1}{2}}$

$$\therefore \ \eta = y\sqrt{\frac{U}{\upsilon x}} \qquad \textbf{(5.31)}$$

The stream function of the flow, $\varphi = \int u\, dy\, |_{x \to const.}$ should increase as δ or $x^{1/2}$ increases and the following nondimensional form:

$$\varphi = \sqrt{\upsilon x U}\, f(\eta) \qquad \textbf{(5.32)}$$

Where f is a function to be determinate. Note from the definition of stream function:

$$u = \frac{\partial \varphi}{\partial y} = \frac{\partial \varphi}{\partial \eta} \times \frac{\partial \eta}{\partial y} = Uf'(\eta) \qquad \textbf{(5.33)}$$

$$\mathrm{v} = -\frac{\partial \varphi}{\partial x} = \frac{1}{2}\sqrt{\frac{\upsilon U}{x}}\,(\eta f' - f) \qquad \textbf{(5.34)}$$

substituting u & v into the boundary layer momentum eq.(5.29) when $\frac{\partial U}{\partial x} = 0$ we get

$ff'' + 2f''' = 0$ Blasius equation **(5.35)**

The consideration manipulation useful student exercise.

Referring to eq.'s (5.33), (5.34) the no – slip conditions

$u_{(x,0)} = \mathrm{v}_{(x,0)} = 0$ and the free stream merge conditions $u_{(x,\infty)} = U$ convert to

$\eta = 0 \rightarrow f'(0) = f(0) = 0$ and $f(\infty) = 1$ for $\eta = \infty$

Equation (5.26) is the celebrated nonlinear Blasius equation for flat – plate flow

5.5.5. The Falkner – Skan Wedge Flows (General Solution)

The most famous family of B.L. similarity solution was discovered by Falkner and Skan and later calculated numerically by Hartree. The solution as we did with the Blasius problem [1, 12].

Then

$$\left.\begin{array}{ll} \eta=\frac{y}{\delta_{(x)}}, & \varphi=U_{(x)}\,\delta_{(x)}\,f_{(\eta)} \\ \delta_{(x)}=\sqrt{\frac{\upsilon x}{U_{(x)}}} & \varphi=\sqrt{U_{(x)}\,\upsilon x}\,f_{(\eta)} \end{array}\right\} \quad (5.36)$$

$$\eta = y\sqrt{\frac{U_{(x)}}{\upsilon x}} \qquad (\eta(x,y))$$

Where $\delta_{(x)}, U_{(x)}, \eta(x,y)$ where the pressure term is not neglected i.e. using eq.(5.29) without neglecting pressure team as done by Blasius eq.(5.30).

Using the above equations (relations) in eq. (5.29) in. x – momentum equation may be transformed to ordinary partial differential eq. of the form.

$$f''' + \alpha f f'' + \beta\left(1 - f'^2\right) = 0 \quad (5.37)$$

$$\text{Where} \quad \left.\begin{array}{l} \beta=\frac{1}{\upsilon}\delta^2_{(x)}\frac{\partial U_x}{\partial x} \\ \alpha=\frac{1}{2\upsilon}U_x\frac{d\delta^2_{(x)}}{\partial x}+\frac{1}{\upsilon}\delta^2_{(x)}\frac{\partial U_x}{\partial x} \end{array}\right\} \quad (5.37\text{a and b})$$

From eq.s (5.37) we get

$$\frac{d}{dx}\left(U_x\delta^2_{(x)}\right) = \upsilon(2\alpha - \beta) \quad (5.38)$$

If $2\alpha - \beta \neq 0$, integral eq. 5.38 we get

$$U_x\delta^2_{(x)} = \upsilon(2\alpha - \beta)x \quad (5.39)$$

Now subtracts both part of eq. (5.37) (a, b) we get

$$(\alpha-\beta)=\frac{1}{2\upsilon}U_x\frac{d\delta_{(x)}^2}{dx}$$

Or

$$\frac{dU_x}{dx}(\alpha-\beta)=\frac{1}{2\upsilon}U_x.\,2\delta_{(x)}\frac{d\delta_{(x)}}{dx}\times\frac{\delta_{(x)}}{\delta_{(x)}}\frac{dU_x}{dx}$$

$$\frac{dU_x}{dx}(\alpha-\beta)=\underbrace{\frac{1}{\upsilon}\delta_{(x)}^2\frac{dU_x}{dx}}_{\beta}\times\frac{U_x}{\delta_{(x)}}\times\frac{d\delta_{(x)}}{dx}$$

$$\frac{dU_x}{dx}(\alpha-\beta)=\beta\cdot\frac{U_x}{\delta_{(x)}}\cdot\frac{d\delta_{(x)}}{dx}$$

$$(\alpha-\beta)\frac{\partial U_x}{U_x}=\beta\cdot\frac{d\delta_{(x)}}{\delta_{(x)}}$$

Integrate both side

$$(\alpha-\beta)\ln U_x=\beta\ln\delta_{(x)}+\ln k$$

$$\therefore\ {U_x}^{\alpha-\beta}=k\delta_{(x)}^{\beta} \qquad \textbf{(5.40)}$$

Eq.(5.40) could be in the form

$$\frac{{U_x}^{\alpha-\beta}}{k^{\frac{1}{\beta}}}=\delta_{(x)} \qquad \textbf{(5.40a)}$$

Now we eliminate $\delta_{(x)}$ between (5.39), (5.40) we get

$$U_x\left(\frac{U_x^{\frac{2(\alpha-\beta)}{\beta}}}{k^{2/\beta}}\right)=\upsilon(2\alpha-\beta)x$$

Or

$$U_x^{\frac{2\alpha-\beta}{\beta}}=k^{2/\beta}\,\upsilon(2\alpha-\beta)x$$

$$U_{(x)}=k^{\frac{2}{2\alpha-\beta}}\,[\upsilon(2\alpha-\beta)x]^{\frac{\beta}{2\alpha-\beta}} \quad \textbf{(5.41)}$$

The velocity profile of Falkner – skan on any surface.

Also for δ_x

$$\delta_{(x)}=\frac{\left\{k^{\frac{2}{2\alpha-\beta}}\,[\upsilon(2\alpha-\beta)x]^{\frac{\beta}{2\alpha-\beta}}\right\}^{\frac{\alpha-\beta}{\beta}}}{k^{1/\beta}}$$

$$\delta_{(x)}=k^{\frac{1}{\beta-2\alpha}}\,[\upsilon(2\alpha-\beta)x]^{\frac{\alpha-\beta}{2\alpha-\beta}} \quad \textbf{(5.42)}$$

To find out the velocity profile as a fuction of α,β

i.e. for $\alpha=1$ and $\beta=\frac{2m}{m+1}$ we get

$$U_{\infty(x)}=k^{m+1}\left[\upsilon\left(\frac{2m}{m+1}\right)x\right]^{m} \quad \textbf{(5.43)}$$

Potential flow velocity, (Fig. **5.8**).

For $\alpha=0$ then

$$U_x=k^{-\frac{2}{\beta}}\,[-\upsilon\beta x]^{-1}=-\frac{k^{-\frac{2}{\beta}}}{\upsilon\beta}x^{-1}$$

Or

$$U_{(x)} = -C \cdot \frac{1}{x} \tag{5.44}$$

For $C > 0$ sink flow or convergent chanal with flat wall

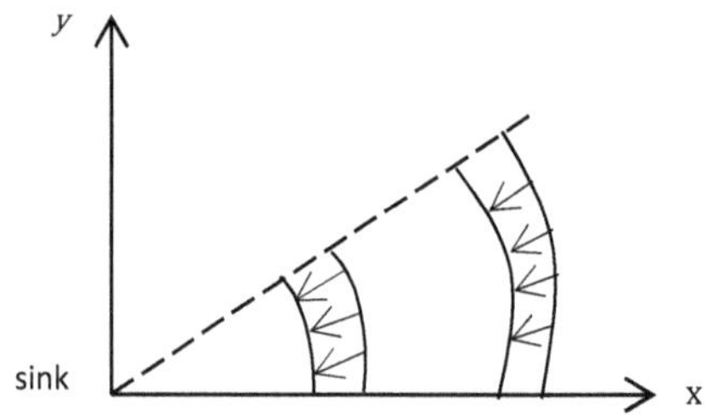

Fig. (5.8). Potential flow velocity.

And $C < 0$ source flow or divergent chanal with flat wall

Case of Steady

1) $\alpha = \frac{1}{2}, \beta = 0 \Rightarrow$ flat plate and eq.(10)

   Falkner – Blasius

2) $\alpha = 1, \beta$ arbitrary flow over a wedge

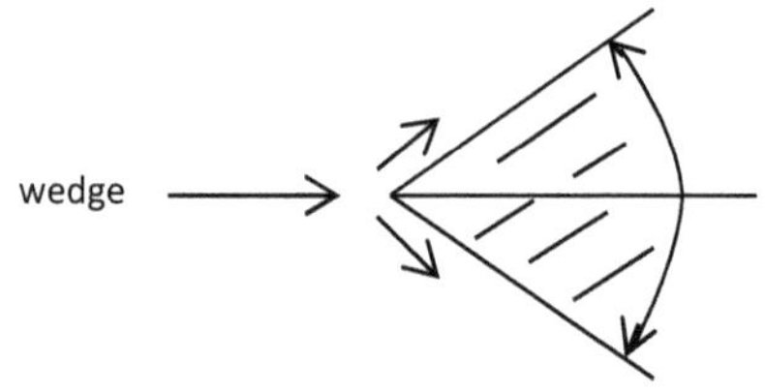

3) $\alpha = 0, \beta = 1$ convergent channal

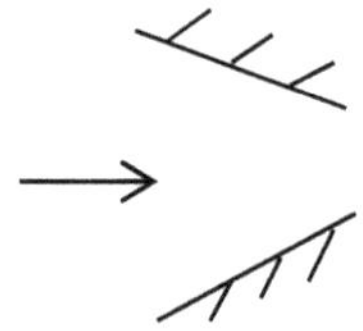

4) $\alpha = 1, \beta = 1$ stagnation point

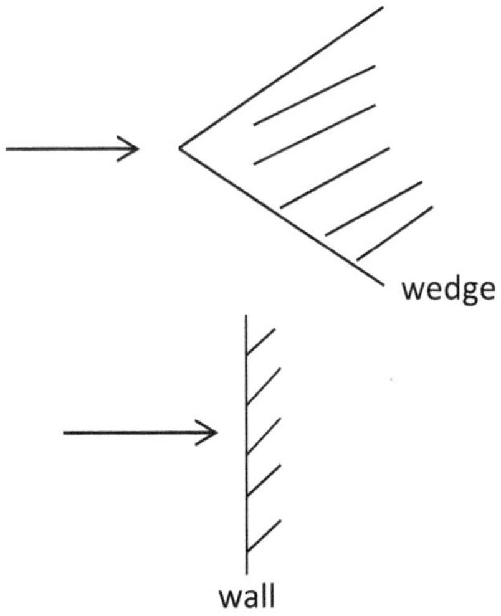

Some examples of Falkner – skan potential flow, (Fig. **5.9**).

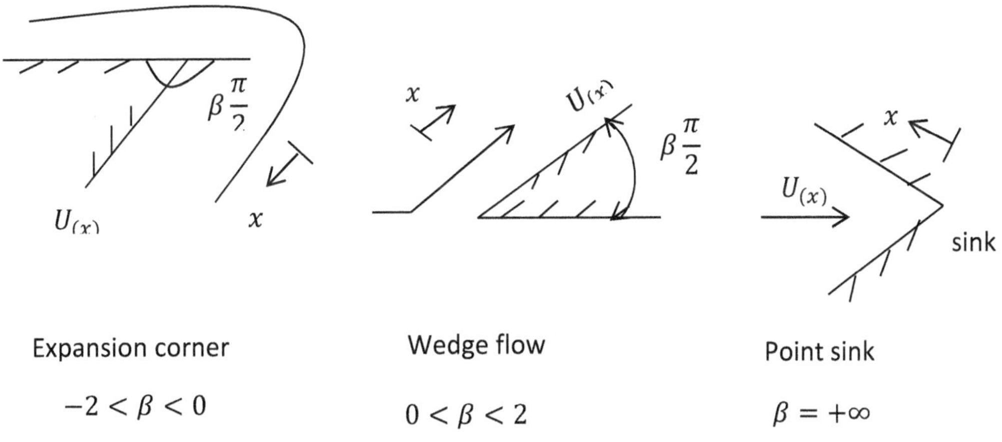

Fig. (5.9). Potential flow examples.

5.5.6. Thwaites Method (Steady Flow)

Both laminar and turbulent theories can be developed from Kaman's general two dimensional B.L. integral relation [1, 5].

$$\tau_w = \rho \frac{\partial}{\partial x} \int_0^\delta (U-u)\, u dy + \rho \frac{dU}{dx} \int_0^\delta (U-u)\, dy \quad \textbf{(5.45)}$$

From the definations of B.L. thickness, momentum thickness and shap factor we get

$$\frac{1}{2} C_f = \frac{\tau_w}{\rho U^2} = \frac{\partial \theta}{\partial x} + \frac{\theta}{U} \frac{\partial U}{\partial x} (2+H)$$

Eq.(5.46) could be integrated to determine θ_x if we correlated C_f and H with momentum thickness.

This has been done by examining typical velocity profiles of laminer and turbulent B.L. flows for various pressure gradients. From experimental shown that shape factor a good indicator of pressure gradient. The higher shape factor H stronger the advarse gradient and spertions occurs approximately.

$$H = \begin{cases} 3.5 & Laminer\ flow \\ 2.5 & Turbulent\ flow \end{cases}$$

A simple and effective methods was developed by Thwaites who found that eq.(5.46) can be rewriting and can be correlated by a single dimensionless momentum thickness λ and shear stress S_λ

Then

$$U \frac{d}{dx} \left(\frac{\theta^2}{\upsilon} \right) = -2 \left[\frac{\theta^2}{\upsilon} \frac{dU}{dx} (H+2) - \frac{\theta}{U} \frac{\tau_w}{\mu} \right] \quad \textbf{(5.47)}$$

And $\lambda = \dfrac{\theta^2}{\upsilon} \cdot \dfrac{dU}{dx}$ pressure gradient factor

$S_\lambda = \dfrac{\theta}{U} \dfrac{\tau_w}{\mu}$ dimensionless shearstress parameter

Subsituting in eq.(5.47) we get

$$U \frac{d}{dx} \left(\frac{\lambda}{dU/dx} \right) = 2 \left[S_{(\lambda)} - \lambda (H+2) \right] \quad \textbf{(5.48)}$$

Thwaites by using a straight – line fit to his correlation was able to integrate eq. (5.48). He assumed that

$$2\left[S_{(\lambda)} - \lambda\,(H+2)\right] = f_{(\lambda)} \quad \textbf{(5.49)}$$

Thwaites procceded to collecte all avalible B.L. soluation from which he was able to plots $f_{(\lambda)}$ versus λ with approximation relation

$$f_{(\lambda)} = 0.45 - 6\lambda$$

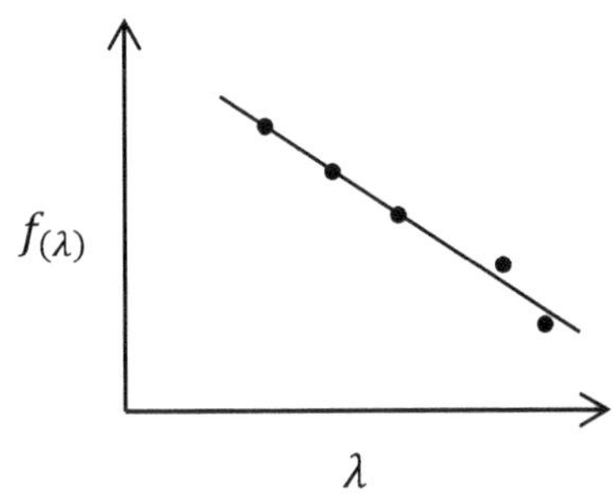

Then substituat for $f_{(\lambda)}$ in eq.(5.48) and intagrate to get

$$\lambda = \frac{0.45}{U^6} \cdot \frac{dU}{dx} \int_0^x U^5\,dx$$

Or

$$\theta^2 = \theta_\circ{}^2 + \frac{0.45\upsilon}{U^6} \int_0^x U^5\,dx \quad \textbf{(5.50)}$$

Where $\theta_\circ$ is the momentum thickness at $x = 0$ (usually taken to be zero). Speration $(C_f = 0)$ was found to occure at a particular value of λ.

At seperation point $\lambda = -0.09$
If U is given, therefore λ is found. Finally, Thwaits correlated of the dimensionless shear stress $S = \tau_w\theta/\mu U$ with λ and his graphed results can curve – fit as follows:

$$S_{(\lambda)} = \frac{\tau_w \, \theta}{\mu \, U} = (\lambda + 0.09)^{0.62}$$

Also when λ is found, plotted from the same collected soluations

$S_{(\lambda)}$ and $H_{(\lambda)}$.

Cebecci and Brand show fitted these functions S and H in the following eq.

$$S_{(\lambda)} = 0.22 + 1.57\lambda - 1.8\lambda^2 \qquad \text{for} \quad 0 < \lambda < 0.1$$

$$= 0.22 + 1.402\lambda + \frac{0.018\lambda}{\lambda + 0.107} \qquad \text{for} \quad -0.1 < \lambda < 0$$

$$H_{(\lambda)} = 2.61 - 3.75\lambda + 5.24\lambda^2 \qquad \text{for} \quad 0 < \lambda < 0.1$$

$$= 2.088 + \frac{0.0731\lambda}{\lambda + 0.14} \qquad \text{for} \quad -0.1 < \lambda < 0$$

Then the problem of thwaites becomes as

$U_{(x)}$ is given get $\Rightarrow \lambda_{(x)}$

Then $S_{(x)}$ and $H_{(x)}$ can be calculated from above relations.

Then as we can get

$$\left.\begin{array}{ll} \lambda = \frac{\theta^2}{\upsilon} \cdot \frac{dU}{dx} & \Rightarrow \theta \\ S = \frac{\theta \, \tau_w}{\mu \, U} & \Rightarrow \tau_w \\ H = \frac{\delta^*}{\theta} & \Rightarrow \delta^* \end{array}\right\} \begin{array}{c} Then\ we\ can\ get\ \theta, \delta^*, \tau_w\ this\ for\ \frac{dP}{dx} \neq 0 \\ condition\ of\ this\ method\ and\ \frac{dP}{dx} = 0 \\ for\ flat\ plate. \end{array}$$

The accuracy of Thwaites method is about ± 5 percent for favorable or mild adverse gradients but may be as much as ± 15 percent near the seperation point. Nevertheless, since the method is an average of many exact solutions, it can be regarded as a best – avaliable one – parameter method. If more accuracy is desired the finite – difference computer method of is recommended.

CHAPTER 6

Turbulent Boundary Layer

Abstract: The greatest number of practical applications the boundary layer is turbulent rather than laminar. Alternatively, more precisely, the bulk of the boundary layer is turbulent, with a thin laminar sublayer near the wall. Physically, turbulence acts to magnify enormously the local, instantaneous gradients of velocity, and thus is greatly augments the viscous stresses and acting in the fluid. The most important practical consequence of this is that the skin – friction for turbulent boundary layer are several orders of magnitude layer than the corresponding values for a laminar boundary layer at the same Reynolds number.

In this chapter, a brief description and derivation of turbulent boundary layer equation. The relation between stress (Reynolds and viscous stresses) and pressure gradients *i.e.* the relative magnitudes of the viscous and Reynolds stresses across a boundary layer. An estimation of laminar sub – layer due to neglecting the Reynolds stresses and the inner region of the turbulent layer by neglecting viscous term. The skin friction coefficient as a function of x and derived their relation with velocity profile along turbulent region.

Keywords: Laminar sub–layer, Reynolds and viscous stresses, Skin friction, Turbulent boundary layer equation.

6.1. TURBULENT B.L. EQUATION

Navier – stokes equations can be used to deyermine turbulent boundary growth and shear stress a flat plate in a manner analogous to the treatment of the laminer boundary layer. Navier – stokes equations flow in x, y, z components as follows [1, 13].

$$\frac{\partial u}{\partial t} + u\frac{\partial u}{\partial x} + \mathrm{v}\frac{\partial u}{\partial y} + w\frac{\partial u}{\partial z} = -\frac{1}{\rho}\frac{\partial P}{\partial x} + \upsilon\left[\frac{\partial^2 u}{\partial x^2} + \frac{\partial^2 u}{\partial y^2} + \frac{\partial^2 u}{\partial z^2}\right] \quad \textbf{(6.1)}$$

$$\frac{\partial \mathrm{v}}{\partial t} + u\frac{\partial \mathrm{v}}{\partial x} + \mathrm{v}\frac{\partial \mathrm{v}}{\partial y} + w\frac{\partial \mathrm{v}}{\partial z} = -\frac{1}{\rho}\frac{\partial P}{\partial y} + \upsilon\left[\frac{\partial^2 \mathrm{v}}{\partial x^2} + \frac{\partial^2 \mathrm{v}}{\partial y^2} + \frac{\partial^2 \mathrm{v}}{\partial z^2}\right] \quad \textbf{(6.2)}$$

Jafar Mehdi Hassan, Riyadh S. Al-Turaihi, Salman Hussien Omran, Laith Jaafer Habeeb, Alamaslamani Ammar Fadhil Shnawa

$$\frac{\partial w}{\partial t}+u\frac{\partial w}{\partial x}+v\frac{\partial w}{\partial y}+w\frac{\partial w}{\partial z}=-\frac{1}{\rho}\frac{\partial P}{\partial z}+\upsilon\left[\frac{\partial^2 w}{\partial x^2}+\frac{\partial^2 w}{\partial y^2}+\frac{\partial^2 w}{\partial z^2}\right] \quad \textbf{(6.3)}$$

$$\text{Continuity eq.} \quad \frac{\partial u}{\partial x}+\frac{\partial v}{\partial y}+\frac{\partial w}{\partial z}=0 \quad \textbf{(6.4)}$$

x, y, z Cartesian coordination

u, v, w the corresponding velocity comp.

ρ, P, υ and t, density, power, kinematic viscosity, time

These eqs. are valid for turbulent boundary layer as well as laminar flow, when we substituting the corresponding velocities as follows:

$$u=\bar{u}+u' \;,\quad v=\bar{v}+v' \;,\quad w=\bar{w}+w' \;,\quad P=\bar{P}+P' \quad \textbf{(6.5)}$$

And we hare the following assumptions

a) $\frac{\partial \bar{u}}{\partial t}=0$ since the main flow is steady

b) $u'\frac{\partial u'}{\partial t}=0$ since $\frac{\partial u'}{\partial t}=\lim_{T_0\to\infty}\frac{1}{T_0}\int_0^{T_0}\frac{\partial u'}{\partial t}dt=\lim_{T_0\to\infty}\frac{u'_{T_0}-u'_{(0)}}{T_0}=0$

c) $u'\frac{\partial \bar{u}}{\partial t}=\overline{u'}\frac{\partial \bar{u}}{\partial t}=0$ since $\overline{u'}=0$ fluctuation vanish

d) $\bar{u}\frac{\partial u'}{\partial t}=\bar{u}\frac{\partial \overline{u'}}{\partial t}=\bar{u}\frac{\partial}{\partial x}(\overline{u'})=0$ since $\overline{u'}=0$

e) $v'\frac{\partial \bar{u}}{\partial y}=w'\frac{\partial \bar{u}}{\partial z}=\bar{v}\frac{\partial u'}{\partial y}=\bar{w}\frac{\partial u'}{\partial z}=0$ similar arguments

f) $\frac{\partial^2 u'}{\partial x^2}=\frac{\partial^2}{\partial x^2}(\overline{u'})=0$ $\quad \overline{u'}=0$

g) Similarly $\frac{\partial^2 u'}{\partial y^2}, \frac{\partial^2 u'}{\partial z^2} \cdots\cdots$

we get the following equations

$$\rho\left(\bar{u}\,\frac{\partial \bar{u}}{\partial x} + \bar{v}\frac{\partial \bar{u}}{\partial y} + \bar{w}\frac{\partial \bar{u}}{\partial z}\right) = -\frac{\partial \bar{P}}{\partial x} + \mu\nabla^2\bar{u} + \frac{\partial \tau_{xx}}{\partial x} + \frac{\partial \tau_{xy}}{\partial y} + \frac{\partial \tau_{xz}}{\partial z} \quad \textbf{(6.6)}$$

Similarly, eq. (2), (3) reduce to

$$\rho\left(\bar{u}\,\frac{\partial \bar{v}}{\partial x} + \bar{v}\frac{\partial \bar{v}}{\partial y} + \bar{w}\frac{\partial \bar{v}}{\partial z}\right) = -\frac{\partial \bar{P}}{\partial y} + \mu\nabla^2\bar{v} + \frac{\partial \tau_{yx}}{\partial x} + \frac{\partial \tau_{yy}}{\partial y} + \frac{\partial \tau_{yz}}{\partial z} \quad \textbf{(6.7)}$$

$$\rho\left(\bar{u}\,\frac{\partial \bar{w}}{\partial x} + \bar{v}\frac{\partial \bar{w}}{\partial y} + \bar{w}\frac{\partial \bar{w}}{\partial z}\right) = -\frac{\partial \bar{P}}{\partial z} + \mu\nabla^2\bar{w} + \frac{\partial \tau_{zx}}{\partial x} + \frac{\partial \tau_{zy}}{\partial y} + \frac{\partial \tau_{zz}}{\partial z} \quad \textbf{(6.8)}$$

Eqs. 6.6 ,6.7, 6.8 are the Reynolds equations for turbulent flow.

Note that they are of the same form as the Navior – stokes equations, apart from the addition of the Reynolds stress terms.

The two – dimensional shear flow equations for steady mean flow were obtaind above (equations 6.6,6.7,6.8) are as follows:

$$\frac{\partial \bar{u}}{\partial x} + \frac{\partial \bar{v}}{\partial y} = 0 \qquad \text{continuity equation} \quad \textbf{(6.9)}$$

$$\bar{u}\,\frac{\partial \bar{u}}{\partial x} + \bar{v}\frac{\partial \bar{u}}{\partial y} = \rho U\frac{dU}{dx} + \frac{\partial \tau}{\partial y} \quad \textbf{(6.10)}$$

Where $-\frac{dP}{dx} = \rho U\frac{dU}{dx} + \frac{\partial \tau}{\partial y}$ [Bernouli equation]

And

$\tau = \mu\frac{\partial \bar{u}}{\partial y} - \rho u'v'$ is the total shear stress.

$\mu \dfrac{\partial \bar{u}}{\partial y}$ is the viscous stress and

$-\rho u'v'$ is Reynolds stress.

6.2. THE RELATIONS BETWEEN STRESSES AND PRESSURE GRADIENTS

In practical, these equations apply to turbulent boundary layer flows. Experimental results provide important guidenlines for the analysis of the equations. For example, the retative magnitudes of the viscous and Reynolds stresses across a boundary layer can be abtained experimentally [1,2]. The following sketches (Fig. **6.1**) show the resulting typical stress distributions for (1) zero pressure grudient. (2) farourable pressure gradient and (3) adverse pressure gradient.

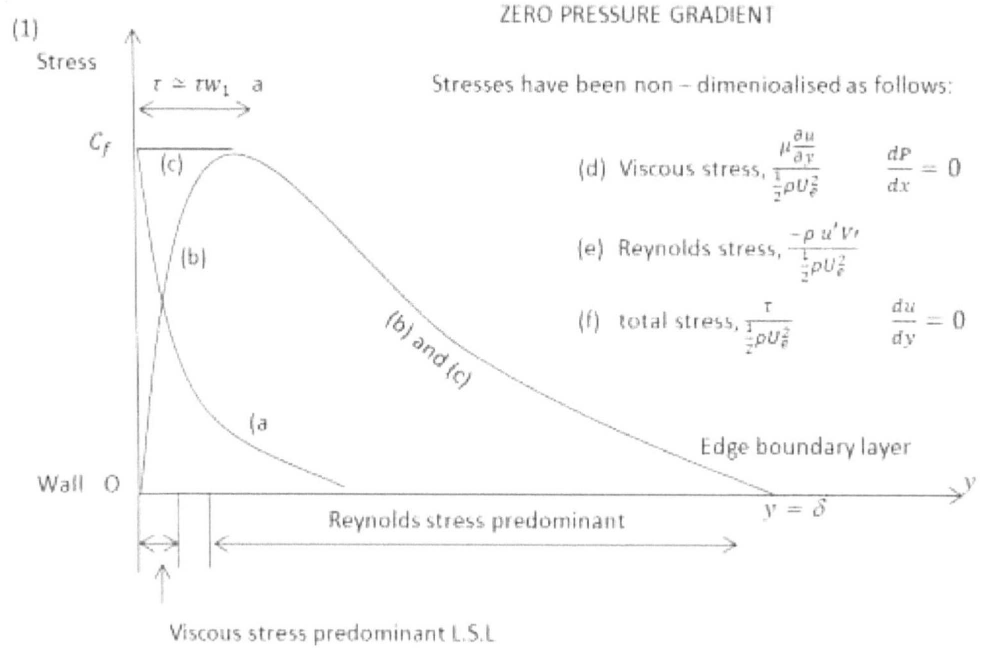

(Fig. 6.1) contd.....

(2)

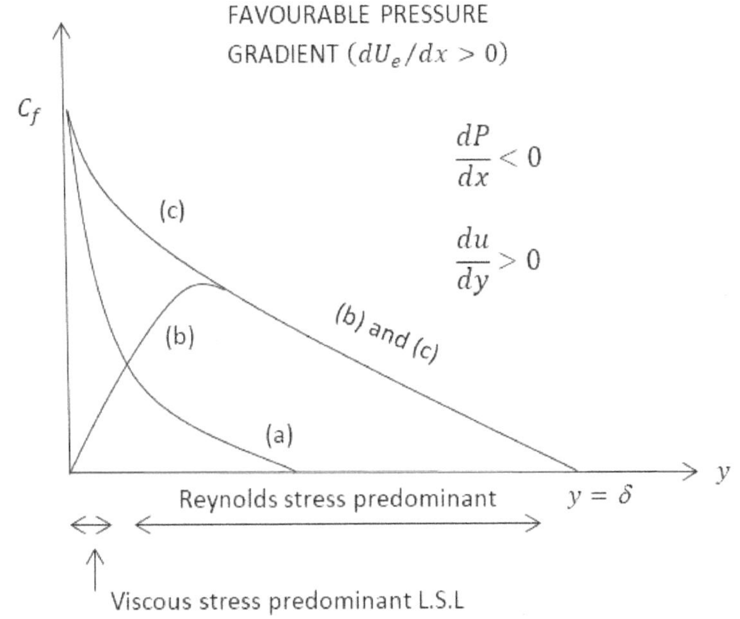

(3)

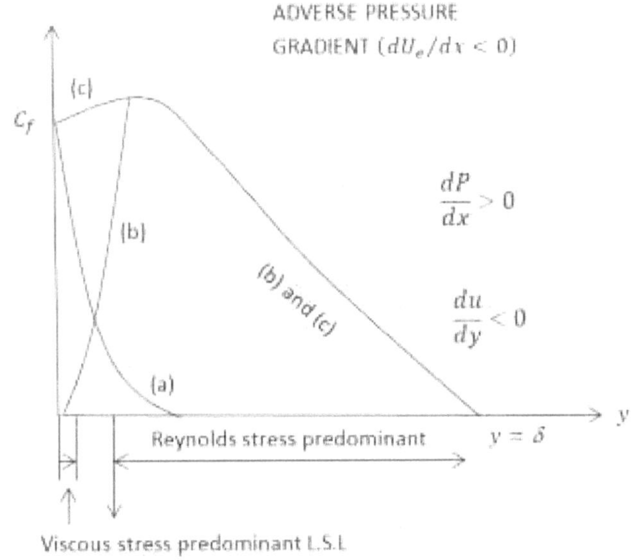

Fig. (6.1). Typical stress distribution for different pressure [2].

In all three case there is a thin layer adjacent to the wall in which the visccus stress is much larger than the Reynolds stresses. This is because

$u' \to 0$ and $v' \to 0$ as $y \to 0$

By no – slip conditions and so the Reynolds stresses,

$-\rho u'v' \to 0$ as $y \to 0$.

This thin region is called the laminar sub – layer. The sketches shows however, that the Reynolds stresses are predominat over the major part of the B.L.

It is also apparent that $\tau = \tau_w$ for small pressure gradients and small values of y. (but extending beyound the laminar sub – layer). This results can also be obtained from equation (6.10) as follows:

If y is sufficiently small then $\bar{u}$ is small an $\bar{v}$ is even smaller, then the convective terms in equation (6.10) can be nelected and so.

$$\frac{\partial \tau}{\partial y} = -\rho U \frac{\partial U}{\partial x}, \quad \text{integration gives}$$

$$\tau = -\rho U \frac{\partial U}{\partial x} y + constant.$$

when $y = 0$ $\tau = \tau_w$ and so

$$\tau = -\rho U \frac{\partial U}{\partial x} + \tau_w$$

But y and $-\rho U \frac{\partial U}{\partial x}$ are small and so $-\rho U \frac{\partial U}{\partial x} y$ is negrneted

$$\therefore \quad \tau = \tau_w \tag{6.11}$$

This result will now be used to obtain $\bar{u}$ in the laminar sub – layer and in the inner regions of the turbulent flow.

6.2.1. Laminar Sub – Layer

Neglecting Reynolds stresses

$$\tau = \mu \frac{\partial \bar{u}}{\partial y} \quad \text{and} \quad \tau = \tau_w$$

$$\therefore \ \mu \frac{\partial \bar{u}}{\partial y} = \tau_w, \quad \text{integration then given}$$

$$\mu \, \bar{u} = \tau_w \, y \tag{6.12}$$

The constant of integration being zero since $\bar{u} = 0$ when $y = 0$

Now $\sqrt{\frac{\tau_w}{\rho}}$ has the dimensions of velocity.

This suggests introducing the " friction velocity " u^*, defined as follows:

$$u^* = \sqrt{\frac{\tau_w}{\rho}} \quad \text{or} \quad {u^*}^2 = \frac{\tau_w}{\rho}$$

Then equation (6.13) gives:

$$\frac{\mu}{\rho} \bar{u} = \frac{\tau_w}{\rho} y$$

$$\text{or} \quad \frac{\bar{u}}{u^*} = \frac{u^* y}{\upsilon} \tag{6.13}$$

the error involved in neglecting the Reynolds stress in the laminer sub – layer can be estimated as follow. Using Taylors theorem for small y.

$$\left(\overline{u'v'}\right)_y = \left(\overline{u'v'}\right)_0 + y\left(\frac{\partial}{\partial y}\overline{u'v'}\right)_0 + \frac{y^2}{2!}\left(\frac{\partial^2}{\partial y^2}\overline{u'v'}\right)_6 + \frac{y^3}{3!}\left(\frac{\partial^3}{\partial y^3}\overline{u'v'}\right)_6 + O(y^4)$$

It can be easily be shown that the first 3 terms of the series are zero (from turbulent flow assumptions) then

$$\left(\overline{u'v'}\right)_y = +\frac{y^3}{3!}\left(\frac{\partial^3}{\partial y^3}\overline{u'v'}\right)_0 + O(y^4)$$

Now

$$\tau = \mu\frac{\partial \bar{u}}{\partial y} - \rho\,\overline{u'v'} = \tau_w$$

$$\therefore \tau_w = \mu\frac{\partial \bar{u}}{\partial y} - \rho\frac{y^3}{6}\left(\frac{\partial^3}{\partial y^3}\overline{u'v'}\right) + O(y^4)$$

Integration and simplifiction give

$$\frac{\bar{u}}{u^*} = \frac{u^* y}{v} + \frac{y^4}{24u^* v}\left(\frac{\partial^3}{\partial y^3}\overline{u'v'}\right)_0 + O(y^5) \quad \text{(Ratta equation)} \qquad \textbf{(6.14)}$$

6.2.2. The Inner Region of the Turbulent Layer

$\tau = -\rho\,\overline{u'v'}$ and $\tau = \tau_w$ neglecting viscous term.

$$\therefore -\rho\,\overline{u'v'} = \tau_w$$

Prandtl's mixing length hypothesis (M.L.H)

$$-\rho\,\overline{u'v'} = \rho\,\ell^2\left(\frac{\partial \bar{u}}{\partial y}\right)^2$$

Since $\frac{\partial \bar{u}}{\partial y} > 0$ in the boundary layer.

Assume that the mixing length ℓ is proportional to y, *i.e.* $\ell = ky$ where k is a constant. Then

$$-\rho\,\overline{u'v'} = \rho\,k^2y^2\left(\frac{\partial \bar{u}}{\partial y}\right)^2 \quad \text{and so}$$

$$\rho\, k^2 y^2 \frac{\partial \bar{u}}{\partial y} = \tau_w \quad i.e. \quad ky\left(\frac{\partial \bar{u}}{\partial y}\right) = \sqrt{\frac{\tau_w}{\rho}} = u^*$$

$$i.e. \quad \frac{1}{u^*}\frac{\partial \bar{u}}{\partial y} = \frac{1}{ky}$$

∴ integrating we get

$$\frac{\bar{u}}{u^*} = \frac{1}{k}\ln y + f(x)$$

In non – dimensional form, this equation becomes:

$$\frac{\bar{u}}{u^*} = A\ln\left(\frac{yu^*}{v}\right) + B \quad \textbf{(6.15)}$$

Where $A = \frac{1}{k}$ and B are constans.

$$\text{or } \frac{\bar{u}}{u^*} = A' \log_{10}\left(\frac{yu^*}{v}\right) + B \quad \textbf{(6.16)}$$

where $A' = \dfrac{A}{\log_{10} e}$

note that $\bar{u} \to 0$ as $y \to 0$. Equation (6.15),(6.16), however, are note valid for very small values of y which correspond to the laminar sub –layer equation (6.13). Holds.

The constants can be obtained by comparing equation (6.16) with experimental results. Thus gives

$A' = 5.8$ and $B = 5.5$

Then $A = A' \log_{10} e = 2.5$ $I.e$ $k = \dfrac{1}{A} = 0.4$

(Von – Karmon's constant).

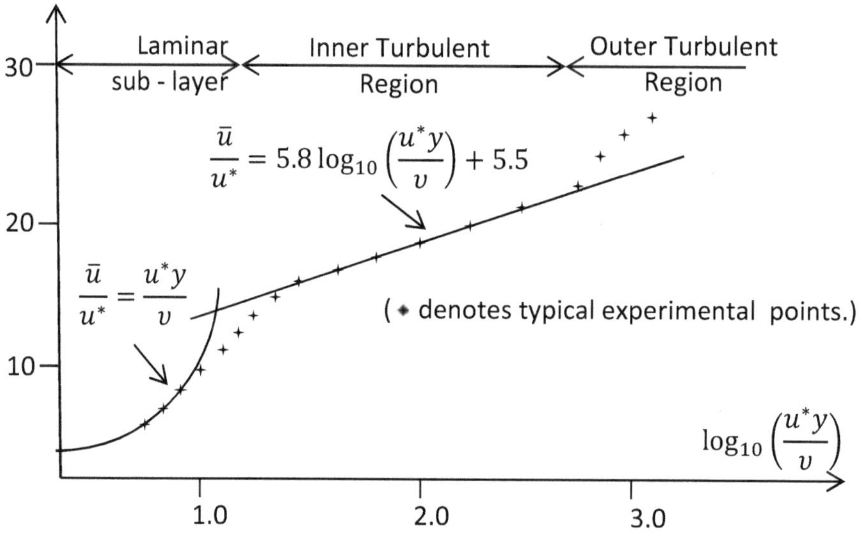

Fig. (6.2). A comparision between logarithmic law and experimental results [1,2].

These the laminar sub – layer extends over the range

$$0 \le \frac{yu^*}{v} < 10$$

and the inner turbulent region extends over the range

$$20 < \frac{yu^*}{v} < 400$$

6.3. THE SKIN FRICTION COEFFICIENT $C_f(x)$

By definition $\tau_w = \frac{1}{2}C_f\rho U^2$

Then

$$\frac{C_f}{2} = \frac{\tau_w}{\rho U^2} \quad \text{and} \quad u^* = \sqrt{\frac{\tau_w}{\rho}} \tag{6.17}$$

$$\therefore \frac{u^*}{U} = \sqrt{\frac{C_f}{2}} \qquad \textbf{(6.18)}$$

1- First, we shall describe an approximate method of estimating C_f at a fixed value of x at which $\bar{u}(y)$ is known, plotted by erases on the graph below. It is assumed that a region of the known velocity profile is represented by the inner region equation obtained before *i.e.*

$$\frac{\bar{u}}{u^*} = 5.8\log_{10}\left(\frac{yu^*}{\upsilon}\right) + 5.5 \qquad \textbf{(6.19)}$$

And multiply each term by $\frac{U}{U}$ we get

$$\frac{\bar{u}}{U} \cdot \frac{U}{u^*} = 5.8\log_{10}\left(\frac{yu^*}{\upsilon} \cdot \frac{U}{U}\right) + 5.5$$

Using equation (1) we get

$$\frac{\bar{u}}{U} = \left[5.8\log_{10}\left(\frac{yU}{\upsilon}\right) + 5.8\log_{10}\left(\frac{u^*}{U}\right) + 5.8\right]\sqrt{\frac{C_f}{2}}$$

$$\therefore \frac{\bar{u}}{U} = \left[5.8\log_{10}\left(\frac{yU}{\upsilon}\right) + 5.8\log_{10}\sqrt{\frac{C_f}{2}} + 5.8\right]\sqrt{\frac{C_f}{2}} \qquad \textbf{(6.20)}$$

This equation is plotted for various assigned values of C_f. The actual value of C_f is determined by the line which is closest to the given profile in Fig. **(6.3)**.

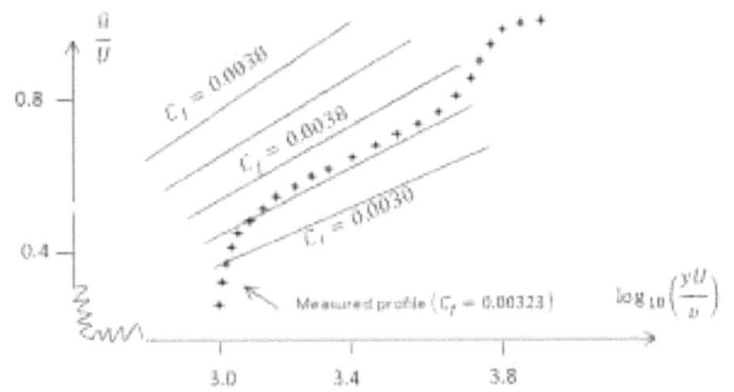

Fig. (6.3). The variation of dimensionless velocity for various assigned value of C_f [2].

2- Since we require C_f as a function of x it is given by the momentum integral equation of boundary layer in terms of U, δ^* and θ where

$$\delta^* = \int_0^\delta \left(1 - \frac{\bar{u}}{U}\right) dy \quad \text{and} \quad \theta = \int_0^\delta \frac{\bar{u}}{U}\left(1 - \frac{\bar{u}}{U}\right) dy \qquad \textbf{(6.21)}$$

Then referring to eq. (5.11) and using the equ (6.21) relation we get

$$\frac{\tau}{\rho U^2} = \frac{\partial \theta}{\partial x} + \frac{\theta}{U}\frac{\partial U}{\partial x}(2 + H) \qquad \textbf{(6.22)}$$

From eq. (1)

$$\frac{C_f}{2} = \frac{\partial \theta}{\partial x} + \frac{\theta}{U}\frac{\partial U}{\partial x}(2 + H) \qquad \textbf{(6.23)}$$

In particular, when $U = constant \ \therefore \ \frac{\partial U}{\partial x} = 0$

Therefore

$$\frac{C_f}{2} = \frac{\partial \theta}{\partial x}\left(\frac{u^*}{U}\right)^2 \qquad \textbf{(6.24)}$$

Now to obtain θ and δ^* we require a relation for $\frac{\bar{u}}{U}$ for $0 \leq y \leq \delta$ *i.e.* valid throughout the boundary layer.

CHAPTER 7

Transition Zone of Boundary Layer

Abstract: Experimental evidence has shown that the transition process in a boundary layer starts locally at a number of points, which are more or less randomly distributed in both stream wise direction near the wall in region of finite stream wise extent at some distance from the leading edge. These transitions probably occur because of local instabilities of the mean basic flow to disturbances.

In this chapter, the transition from laminar to turbulent flow zone discussed physically and theoretically. The stability of flow and their relation with disturbance of flow presented in brief way.

The boundary layer velocity profile and all the related divination are derived by assuming that the momentum thickness will remain constant across the transition position.

Finally, the methods of boundary layer control to prevent separation in order to reduce and attain nigh lift described.

Keywords: Boundary layer control, Momentum thickness, Stability of flow, Transition flow zone.

7.1. TRANSITION AND TURBLANCE

For laminar flow the path of each fluid is a straight line parallel to paths. It is assumed that the plates are perfectly smooth and that the oncoming flow is perfectly laminar. These assumptions, however, are physically unrealistic. No boundary can be perfectly smooth and the oncoming flow may contain disturbances due to dirt or eddies [1, 2].

1. These small disturbances are, under certain condition, rectified or damped out so that the motion remains, laminar. The flow is then termed Stable.
2. On the other hand, the disturbance can produce further disturbance so that the flow becomes increasingly disorganized. In this case the flow is termed unstable and the resulting disorganized flow called turbulent.

Jafar Mehdi Hassan, Riyadh S. Al-Turaihi, Salman Hussien Omran, Laith Jaafer Habeeb, Alamaslamani Ammar Fadhil Shnawa

Transition from laminar to turbulent flow was first observed by O. Reynolds (1880) for flow through a pipe. He observed that the type of flow depends on the Reynolds No. as follow:

i. Low Re, Laminar.
ii. $Re = Re_{crit} \approx 2300$.
iii. $Re > Re_{crit}$, Turbulent.

Re_{crit}. Depends on the initial disturbance in the fluid, the smoothness of the bell - mouth and mechanical vibration. By reducing disturbances, transition Reynolds number of about 20,000 have been attained. On the other hand, for no initial disturbances does Re_{crit}. full below about 2,000 (minimum Re_{crit}).

The rate of change of disturbance energy of a fluid particle depends on:

i. The rate of work done by the additional stress on the bounding surface of the particle due to disturbance velocities (say, A). this may either increase or decrease the disturbance energy.
ii. The rate at which the energy of the disturbing motion is dissipated by viscosity (say. B). this will tend to decrease the disturbance energy of the particle.

If Re is sufficiently small, then B > A and so the disturbances are damp out by viscosity, resulting in stable flow.

7.2. STABILITY ANALYSIS

To analysis of the flow it is sufficient to consider the behavior of the wave number α and the wave length of the disturbance $L = \frac{2\pi}{\alpha}$. As a function of Reynolds No. The typical neutral stability curve for (Re, $\bar{a}$) plane illustrated below. $\bar{\alpha} = L\alpha$ (Fig. **7.1**).

1. If $Re < Re_{crit}$ the flow is stable for all values of $\bar{a}$ (*i.e.* all disturbance). The viscous force damps out all disturbances.
2. If the wave number $\bar{\alpha}$ is large (*i.e.* small wave length), the flow is for all Re. the disturbance velocities have large gradients (*e.g.* $\frac{\partial u}{\partial y}$ large) causing the disturbance to damp by viscousity.

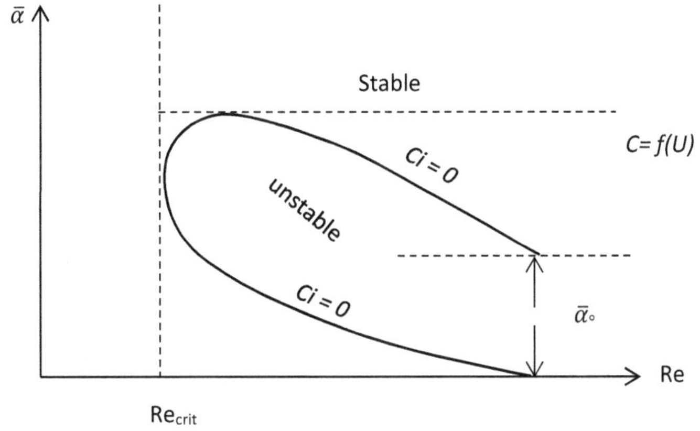

Fig. (7.1). Typical neutral stability curve [2].

In this context, a boundary layer flow may be regarded as unidirectional because change along the boundary are much smaller than velocity changes normal to the boundary. To determine the stability of the laminar B.L. flow at a particular station $x = x_0$ along the boundary, $y = 0$, substituting the laminar velocity profile at $x = x_0$ in place of $U(y)$. this approach has the following wetness.

Stability curves depends only on the local velocity profile and so to not take account of upstream influence (Fig. **7.2**).

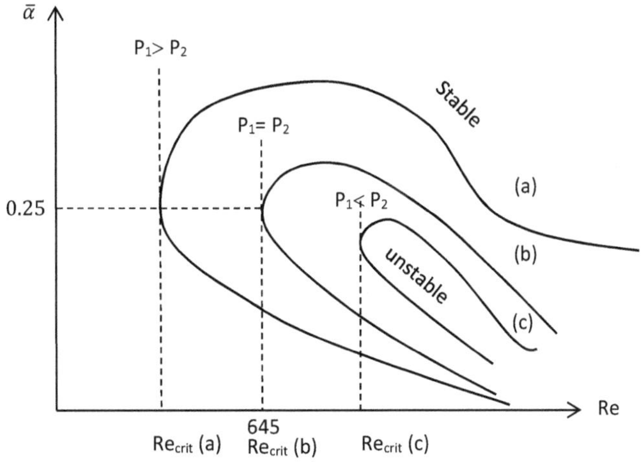

Fig. (7.2). Typical neutral stability curves for [2].

a) Adverse pressure gradient and point of inflexion of the velocity profile.
b) Zero pressure gradient.
c) Favorable pressure gradient (change the pressure gradient to get the stability).

7.3. TRANSITION ZONE

7.3.1. Conditions at Transition

It is usually assumed that the transition from laminar to turbulent flow within the B.L. occurs instantaneously. This is obviously not exactly true, but observation of the transition process does indicate that the transition region (stream wise distance) is very small, so that as a first approximation the assumption is reasonably justified. The assumption for this zone that the momentum per unit time in the flow is the same from upstream till downstream of the transition, thus [14]:

$\theta_{Lt} = \theta_{Tt}$ $\qquad$ t = transition, L = Laminar, T = Turbulent

The ratio of the turbulent to laminar layer 99% thickness is then given directly by

$$\frac{\delta_{Tt}}{\delta_{Lt}} = \frac{{}_L\int_0^1 f(\eta)(1-f(\eta))d\eta}{{}_T\int_0^1 f(\eta)(1-f(\eta))d\eta}$$

When $\quad \theta = \delta \int_0^1 f(\eta)(1-f(\eta))d\eta$

for $\left.\begin{array}{l} f(\eta)_L = \sin\frac{\pi}{2}\eta \\ f(\eta)_T = \eta^{1/7} \end{array}\right\}$ This leads that $\quad \dfrac{\delta_{Tt}}{\delta_{Lt}} = 1.4$

This indicates that on flat plate. The B.L. increases in thickness by 40% at transition.

7.3.2. Mixed B.L. Flow on a Flat Plate with Zero Pressure Gradient

A B.L. will begin to develop thickening with distance downstream as shown in Fig. **(7.3)**, unit transition to turbulence occur at some Re. No. $Re_t = U_e X_e/v$. At thickness increase suddenly from δ_{Lt} in laminar to δ_{Tt} in to turbulent layer and given the relation

$$\delta_{Tt} = \frac{0.383\, X_{Tt}}{(Re_x)_{Tt}^{0.2}}$$ for seventh root profile.

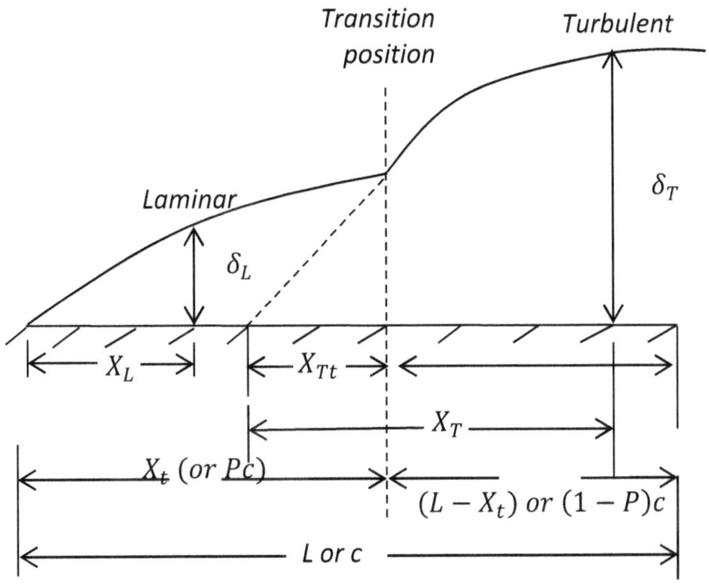

Fig. (7.3). The symbols employed to denote the various physical dimensions used [14].

The drag coefficient C_D for one side of the plate length L may be found by adding the surface fraction force unit width for the laminar large of length X_t to that the turbulent layer of length $(L - X_t)$, and dividing by $\frac{1}{2}\rho U_1^2 L$, where L is here the wetted surface area substituting for δ_{Tt}, then given X_{Tt} in terms of U_1, υ and Re_t then

$$X_{Tt} = 35.6 \frac{\upsilon}{U_1} {Re_t}^{5/8}$$

Referring to previous figure shows that the total effective length of turbulent layer is therefore

$L - X_t + X_{Tt}$ then

$$F = \int_{0}^{L-X_t+X_{Tt}} \tau_w \, dx = \frac{1}{2}\rho U_1^2 \int_{0}^{L-X_t+X_{Tt}} C_f$$

Where $C_f = \frac{\tau_w}{\frac{1}{2}\rho U_1^2} = \frac{0.0595}{(R_{ex})^{1/5}}$ using seventh order law

$$= 0.0595 \left(\frac{\upsilon}{U_1}\right)^{1/5} X^{-1/5}$$

Thus

$$F = \frac{1}{2}\rho U_1^2 \times 0.0595 \left(\frac{\upsilon}{U_1}\right)^{1/5} \times \frac{5}{4} \left[X^{4/5}\right]^{L-X_t+X_{Tt}}$$

Then

$$C_D = \frac{F}{\frac{1}{2}\rho U_1^2 L} = 0.0744 \left(\frac{\upsilon}{U_1}\right)^{1/5} \times \frac{(L - X_t + X_{Tt})}{L}$$

$$C_D = \frac{0.0744}{Re}\left(Re - Re_t + 35.6\, Re_t^{5/8}\right)^{4/5}$$

Per unit width – working in terms of $Re_{t,}$ the transition position is given by

$$X_t = \frac{\upsilon}{U_1} Re_t$$

The laminar layer thickness at transition is the obtained for sine profile *i.e.* $f(\eta) = \sin\frac{\pi}{2}\eta$

$$\delta_{L_t} = \frac{4.79 X_t}{(Re_t)^{1/2}} = 4.79 \left(\frac{\upsilon}{U_1}\right)^{1/2} X_t^{1/2}$$

Substituting for X_t from the previous eq.

$$\delta_{L_t} = 4.79 \frac{\upsilon}{U_1} (Re_t)^{1/2}$$

The corresponding turbulent layer thickness at transition zone

$$\delta_{T_t} = 1.4\, \delta_{L_t} = 6.71 \frac{\upsilon}{U_1} (Re_t)^{1/2}$$

The equivalent length of turbulent layer (X_{T_t}) give this thickness is obtained from

$$\delta_{T_t} = \frac{0.383\, X_{T_t}}{(Re_x)_{T_t}^{1/5}} = 0.383 \left(\frac{\upsilon}{U_1}\right)^{1/5} X_{T_t}^{\ 4/5}$$

Leading to

$$X_{T_t} = 3.3 \left(\frac{U_1}{\upsilon}\right)^{1/4} \delta_{T_t}^{\ 5/4}$$

The expression enables the curves C_f or C_D for the flat plate to be plotted against plate Reynolds No. $Re = (U_1 \upsilon / L)$ for a known value of the transition Re. No. Two such curves for extreme value of Re. of 3×10^5 and 3×10^6 are plotted as follows (Fig. **7.4**).

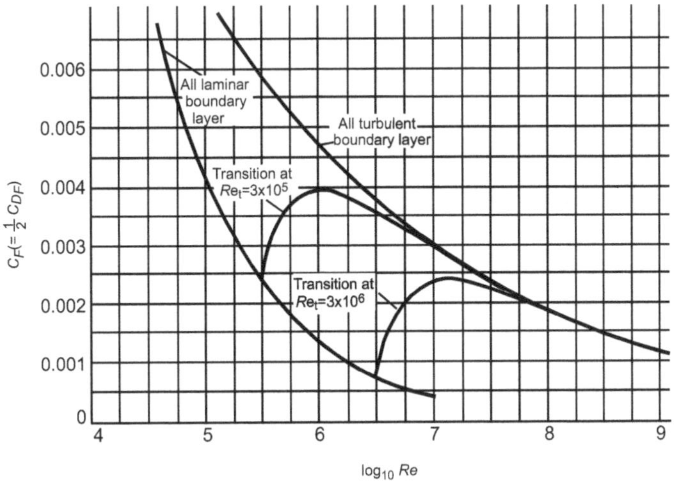

Fig. (7.4). Two – dimensional surface friction drag coefficients for a flat plate. Here Re = plate Reynolds number, *i.e.* $U_\infty L/\upsilon$, $Re_t =$ transition Reynolds number, *i.e.* $U_\infty X_t/\upsilon_1$; $C_f = F/\frac{1}{2}\rho U_\infty^2\, L$; F = skin friction force per (unit width) [14].

7.4. METHODS OF BOUNDARY – LAYER CONTROL

In actual application it is often necessary to prevent separation in order to reduce and to attain high lift several methods of controlling the boundary layer have been developed experimentally, and also on the basis theoretical consideration [4, 15]. Then can be classified?

The derivation of drag coefficients and drag force is useful for student exercise.

7.4.1. Method of the Solid Wall

To avoiding separation is to attempt to prevent the formation of a boundary layer since a boundary – layer owes its existence to the difference between the velocity of the fluid and that of the solid wall, it is possible to eliminate the formation of a boundary – layer by attempting to suppress that difference *i.e.* by causing the solid wall to move with the stream. The simplest way of achieving such a result involves the rotation of a circular cylinder. Fig (**7.5**) shows the flow pattern, which exists a boot of rotating cylinder placed in a stream at right angles to the axis. On the upper sider, where the flow and the cylinder move in the same direction, separation is completely eliminated.

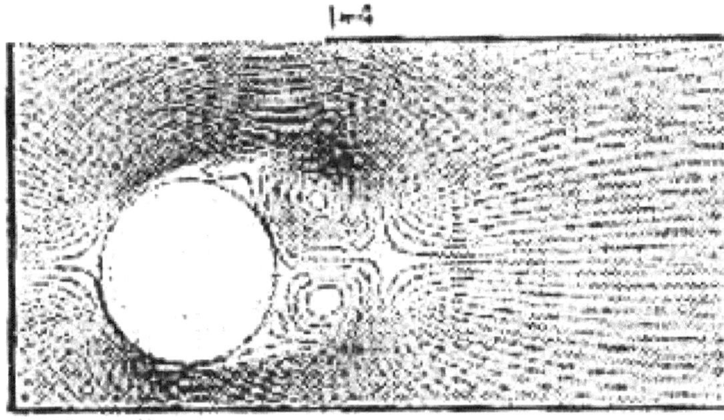

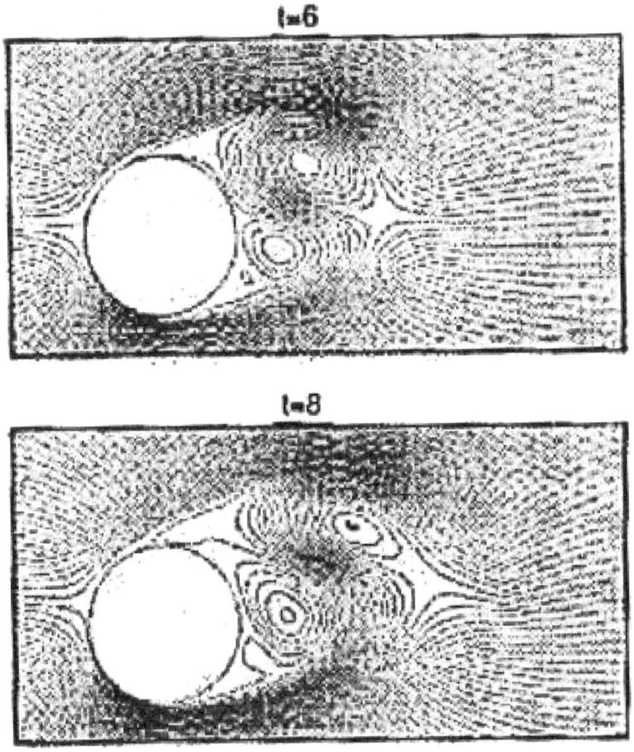

Fig. (7.5). Flow past a rotating cylinder [16].

7.4.2. Acceleration of the Boundary Layer (Blowing)

To preventing separation, supplying additional energy to the particles of fluid which are being regarded in the boundary layer. This result can be achieved by discharging fluid from the body with aid of special blower (Fig. **7.6a**) or by deriving the required energy directly from the main stream, (Fig. **7.6b**). In this way, it is possible to relegate separation. In this way, the gain in the lift force is seen to be very considerable.

7.4.3. Suction

The effect of suction consists in the removal of decelerated fluid particles from the boundary layer before they are given a chance to cause separation, (Fig. **7.6c**). A new boundary layer which is again capable of overcoming a certain adverse pressure gradient is allowed to form in the region behind the slit. With a suitable arrangement of the slits and under favorable conditions, separation can be prevented completely. Simultaneously, the amount of pressure drag is greatly reduced owing to the absence of separation. The application of suction was later widely used in the design of aircraft wings (Fig. **7.7**).

7.4.4. Injection of Different Gas

Which is different from that in the external stream through a porous wall into the boundary layer reduces the rate at which heat is exchanged between the wall and the stream. This is the most important one of the effects produced this way, and for this reason, an arrangement of this kind is often used to provide thermal protection at high supersonic velocities. Injection creates a gaseous mixture in the boundary layer, and to the processes of momentum and heat transfer there is added the process of mass transfer by diffusion.

7.4.5. Prevention of Transition by the Provision of Suitable Shapes

Transition from laminar to turbulent can also be delayed by the use of suitably shaped bodies. The object as in the case of suction is to reduce frictional dray by causing the point of transition to move downstream. It has been established that the location of the point of transition in the boundary layer is strongly influenced by the pressure gradient in the external stream. With a decrease in pressure, transition occurs at much higher Reynolds numbers than pressure increase. A decrease in pressure has highly stabilizing on the boundary layer, and the opposite is true of an increase in pressure along the stream. This circumstance is utilized in modern low – drag airfoils. The desired result is achieved by displacing the suction of maximum thickness for rearwards. In this manner, a large portion of the aerofoil remains under the influence of a pressure, which decreases downstream, and a laminar boundary layer is maintained.

7.4.6. Cooling of the Wall

In a certain range of supersonic Mach numbers, it is possible completely to stabilize the boundary layer by the application of cooling at the wall. Cooling can also be applied in order to reduce the thickness of the boundary layer, and this possibility may become important, *e.g.*, when gases of very low density are made to flow through the nozzles of wind tunnels, because otherwise the very thick boundary layers would unacceptably reduce the useful area of the test section.

The method of boundary layer control of suction, together with the prevention of transition on laminar airfoils, have the great practical importance among all the methods discussed previously. For this reason, various mathematical method for the calculation pf the influence of suction on boundary layer flow have been developed.

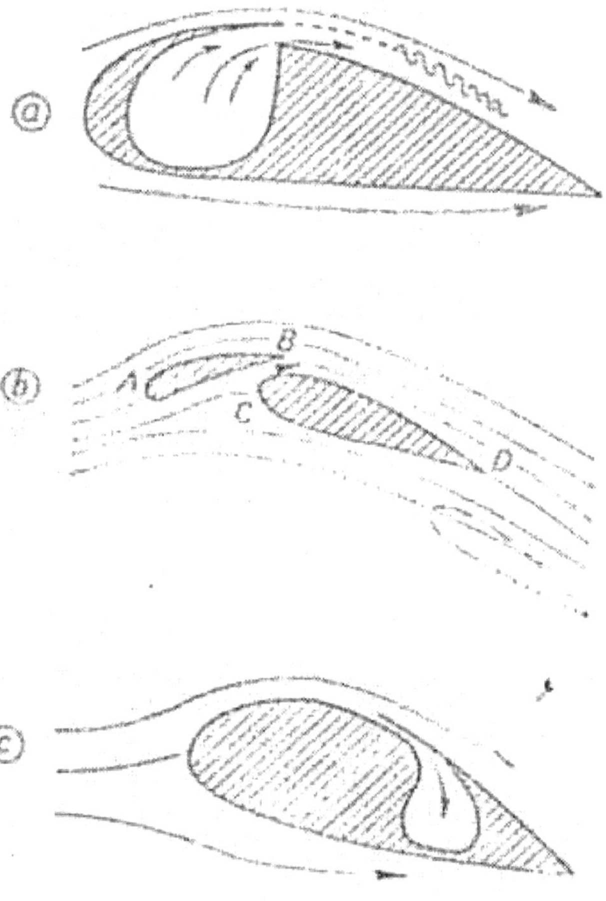

Fig. (7.6). Different arrangcments for boundary – layer control [4]. **(a)** Discharge of fluid, **(b)** Slotted wing, **(c)** Suction.

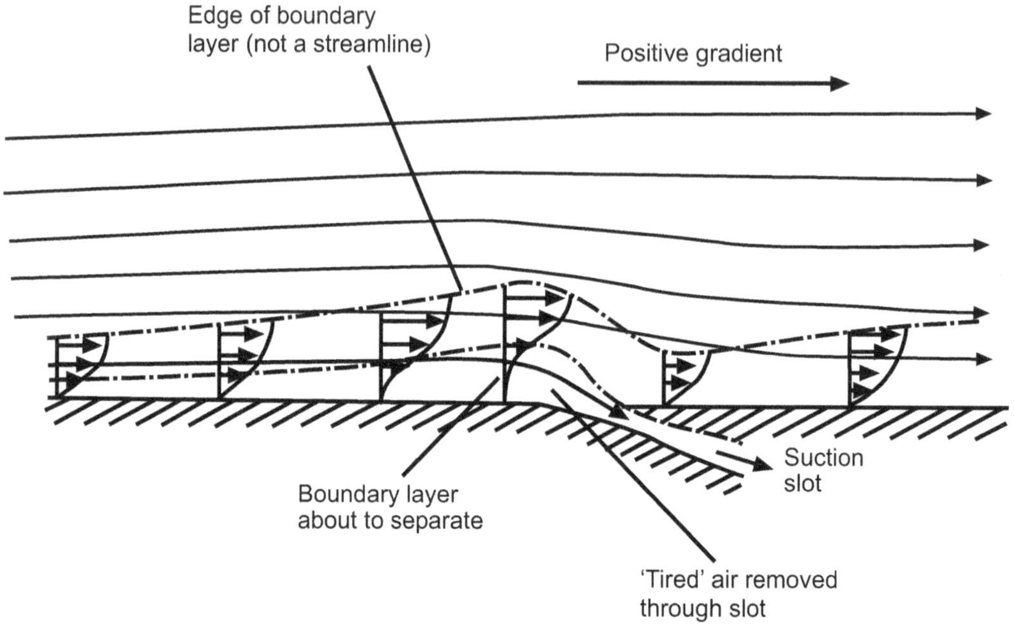

Fig. (7.7). Illustration the suction slot to prevent separation [16].

SOLVED PROBLEMS

Q.1). One method of B.L. control is to use suction as shown in sketch. Clearly this help to limit B.L. growth and delay separation. Consider the suction velocity (V_0) at the plate surface to be uniform. Assume that ($V_0/U_\infty \ll 1$) and that lead to the B.L. thickness and velocity profile both to be independent of x.

For steady laminar B.L. over flat plat show that:

$$\frac{u}{U_\infty} = 1 - e^{-\frac{V_0 y}{\upsilon}} \qquad \frac{\delta^* V_0}{\upsilon} = 1 \qquad C_f = \frac{2V_0}{U_\infty}$$

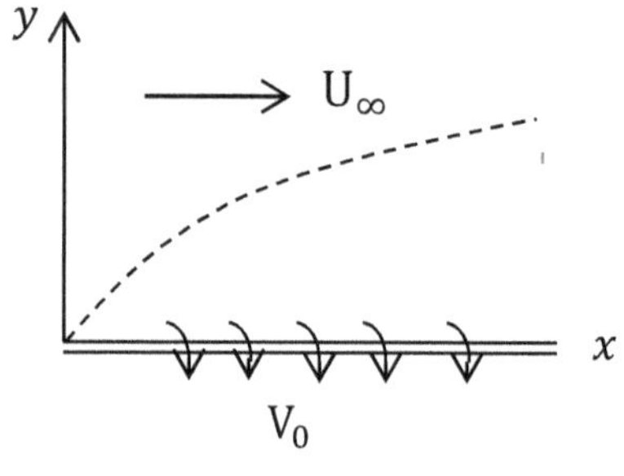

Solution:

For 2-D laminar B.L. the order of magnitude analysis reduces N.S.E. to

N.S.E. $u\frac{\partial u}{\partial x} + \mathrm{v}\frac{\partial u}{\partial y} = \upsilon\frac{\partial^2 u}{\partial y^2}$ ……**(1)** neglecting pressure variation

(free stream)

Continuity eq. $\frac{\partial u}{\partial x} + \frac{\partial \mathrm{v}}{\partial y} = 0$ ……**(2)**

Given that velocity profile independent of $x \Rightarrow u'' u(y) \quad i.e \frac{\partial u}{\partial x} = 0$

Eq. **(2)** becomes $\frac{\partial \mathrm{v}}{\partial y} = 0 \quad \Rightarrow \mathrm{v} = \mathrm{v}(x)$ only

at $y = 0 \Rightarrow \mathrm{v} = -\mathrm{v}_0$ along the plate (always)

eq. (1) becomes

$$-\mathrm{v}_0\frac{\partial u}{\partial y} = \upsilon\frac{\partial^2 u}{\partial y^2} \qquad \therefore \; -\frac{\mathrm{v}_0}{\upsilon} = \frac{\partial^2 u/\partial y^2}{\partial u/\partial y}$$

Integrate both side with respect to y

$$\int -\frac{\mathrm{v}_0}{\upsilon}\,dy = \int \frac{\partial^2 u/\partial y^2}{\partial u/\partial y}\,dy$$

$$-\frac{\mathrm{v}_0}{\upsilon}y = \ln\frac{du}{dy} + \ln c_1 = \ln c_1\frac{du}{dy}$$

$$\text{or} \quad c_1\frac{du}{dy} = e^{-\frac{\mathrm{v}_0}{\upsilon}y}$$

integration again $c_1\int\frac{du}{dy}dy = \int e^{-\frac{\mathrm{v}_0}{\upsilon}y}\,dy$

$$c_1 u = -\frac{\upsilon}{v_0} e^{-\frac{v_0}{\upsilon} y} + c_2$$

B.C. at $y = 0$ $u = 0$

at $y = \infty$ $u = U_\infty$

$$\therefore \quad c_2 = \frac{\upsilon}{v_0} \qquad c_1 U_\infty = -\frac{\upsilon}{v_0} . \frac{1}{e^{\frac{v_0}{\upsilon}\infty}} + c_2$$

$$\therefore \quad c_2 = \frac{\upsilon}{v_0} \quad \text{or} \quad c_1 = \frac{\upsilon}{v_0 U_\infty}$$

$\therefore$ The velocity profile is

$$\frac{\upsilon}{v_0 U_\infty} . u = -\frac{\upsilon}{v_0} . e^{-\frac{v_0}{\upsilon} y} + \frac{\upsilon}{v_0}$$

$$\frac{u}{U_\infty} = 1 - e^{-\frac{v_0}{\upsilon} y}$$

The displacement thickness δ^* givain by

$$\delta^* = \int_0^\infty \left(1 - \frac{u}{U_\infty}\right) dy = \int_0^\infty \left(1 - 1 + e^{-\frac{v_0}{\upsilon} y}\right) dy$$

$$\delta^* = \int_0^\infty e^{-\frac{v_0}{\upsilon} y}\, dy = \left[-\frac{\upsilon}{v_0} . e^{-\frac{v_0}{\upsilon} y}\right]_0^\infty$$

$$= -\frac{\upsilon}{v_0}\left[\frac{1}{e^\infty} - \frac{1}{e^0}\right] = \frac{\upsilon}{v_0}$$

$$\therefore \quad \delta^* = \frac{\upsilon}{v_0}$$

$$c_f = \frac{\tau_0}{\frac{1}{2}\rho U_\infty^{\ 2}} \Rightarrow \tau_0 = \mu \left.\frac{du}{dy}\right|_{y=0} = \mu \frac{v_0}{\upsilon} . U_\infty . \frac{1}{e^{\frac{v_0}{\upsilon} y}} = \mu \frac{v_0}{\upsilon} . U_\infty$$

$$\therefore \quad c_f = \frac{2\,\mu\,\mathrm{v}_0 \mathrm{U}_\infty}{\rho \mathrm{U}_\infty{}^2 \upsilon} = \frac{2\,\upsilon\,\mathrm{v}_0}{\upsilon\,\mathrm{U}_\infty} = \frac{2\,\mathrm{v}_0}{\mathrm{U}_\infty}$$

Q.2). A flow regulator consists of passages (squared section)(2.5 × 2.5) cm^2 and length 15 cm. Water enters at (4 m/s). A boundary layer is forced on the plate of these passages, calculate the displacement thickness at exit section. Calculate the pressure drop due to flow of water through these passage knowing that:

$$\frac{u}{U} = \left(\frac{y}{\delta}\right)^{1/8}, \upsilon = 1.25^{*}10^{-6}\ m^2/s, \text{Turbulent B. L. and}$$

$$\tau_w = 0.0228\, \rho U^2 Re_\delta{}^{-\frac{1}{4}}$$

Solution:

$$\tau_w = \rho\, U_\infty^2 \frac{\partial \delta}{\partial x} \int_0^\delta \frac{u}{\mathrm{U}_\infty}\left(1 - \frac{u}{\mathrm{U}_\infty}\right) dy$$

$$\text{let } \frac{u}{\mathrm{U}_\infty} = f(\eta) = \left(\frac{y}{\delta}\right)^{1/8} = \eta^{1/8}$$

$$\text{or } \tau_w = \rho\, U_\infty^2 A \frac{\partial \delta}{\partial x}$$

$$A = \int_0^1 f(\eta)\,(1 - f(\eta))\, d\eta$$

$$= \int_0^1 \eta^{1/8}\left(1 - \eta^{1/8}\right) d\eta = \int_0^1 \left(\eta^{1/8} - \eta^{2/8}\right) d\eta$$

$$A = \left[\frac{8}{9}\,\eta^{\frac{9}{8}} - \frac{8}{10}\,\eta^{\frac{10}{8}}\right]_0^1 = \left[\frac{8}{9} - \frac{8}{10}\right]$$

$$A = \frac{8}{90} = 0.0889$$

$$\tau_w = 0.0889\, \rho\, U_\infty^2 \frac{\partial \delta}{\partial x} = 0.0228\, \rho\, U_\infty^2\, Re_\delta^{-\frac{1}{4}}$$

$$3.899\, \frac{\partial \delta}{\partial x} = \left(\frac{v}{\mathrm{U}\delta}\right)^{\frac{1}{4}}$$

$$\int \delta^{\frac{1}{4}}\, d\delta = \left(\frac{v}{\mathrm{U}}\right)^{\frac{1}{4}} \cdot \frac{1}{3.899} \int_0^x dx$$

$$\frac{4}{5}\, \delta^{\frac{5}{4}} = \left(\frac{v}{\mathrm{U}}\right)^{\frac{1}{4}} \cdot \frac{1}{3.899}\, x$$

$$\delta^{\frac{5}{4}} = 0.32 \left(\frac{v}{\mathrm{U}}\right)^{\frac{1}{4}} x$$

$$at\;\; x = L = 15\, cm = 0.15\, m$$

$$\delta^{\frac{5}{4}} = 0.32 \times 0.15 \left(\frac{v}{\mathrm{U}}\right)^{\frac{1}{4}}$$

$$= 0.0481 \left(\frac{v}{\mathrm{U}}\right)^{\frac{1}{5}}$$

$$\delta = (0.0481)^{\frac{1}{5}} \left(\frac{v}{\mathrm{U}}\right)^{\frac{1}{5}} = 0.027\, m$$

$$\delta^* = \delta \int_0^1 \left(1 - f(\eta)\right) d\eta = \delta \left[\eta - \frac{8}{9}\eta^{\frac{9}{8}}\right]_0^1$$

$$\delta^* = \delta \left[1 - \frac{8}{9}\right] = \frac{1}{9}\delta = 0.003\, m$$

$$A_1 U_1 = A_2 U_2$$

$(2.5 \times 2.5)10^{-4} \times 4 = (2.5 \times 10^{-2} - 2\delta^*) \times 2.5 \times 10^{-2} U_2$

$U_2 = 5.264\ m/s$

$\therefore \quad \frac{\Delta P}{\gamma} = \frac{U_2^{\ 2} - U_1^{\ 2}}{2g} = 0.6$

$\Delta P = 5.85\ kpa$

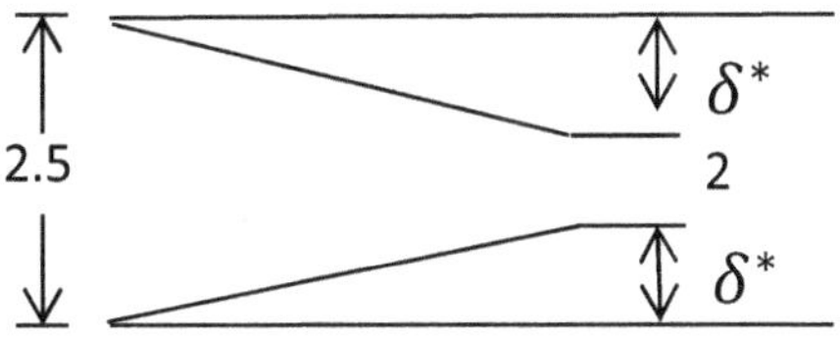

Q.3). The integral B.L. Eq. Written as follows:

$$\frac{C_f}{2} = \frac{\partial \theta}{\partial x} + \frac{\theta}{U}\frac{\partial U}{\partial x}(2 + H)$$

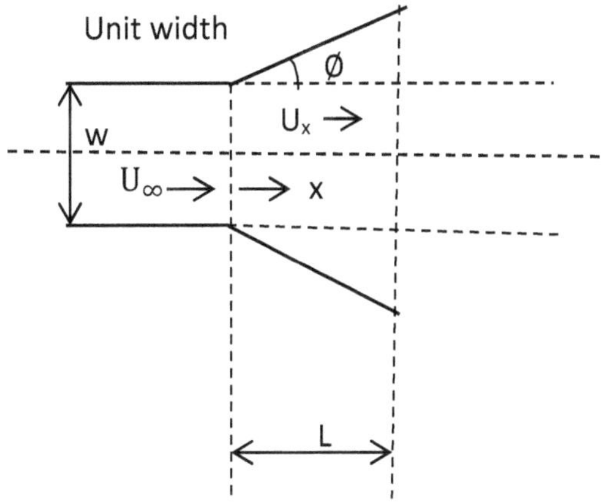

Where: $H = \frac{\delta^*}{\theta}$ Shape factor

$\lambda = \frac{\theta^2}{\upsilon}\frac{\partial U}{\partial x}$ Pressure gradient factor

$S_{(\lambda)} = \frac{\theta \tau_w}{\mu U}$ Shear Stress parameter

Using Thwaites straight-line correlation

$$f_{(\lambda)} = 0.45 - 6\lambda \quad \text{and} \quad f_{(\lambda)} = 2\left[S_{(\lambda)} - \lambda(H + 2)\right]$$

a) Show that $\theta^2 = \frac{0.45\,\upsilon}{U^6}\int_0^x U^5 . dx$

b) For the flat-wall diffuser shown in figure above find the velocity profile then evaluate the pressure gradient factor λ .

c) For L = 2w, find the angle ∅ at the suppression point ratio when $\lambda_{sep} = -0.09$.

Solution:

$$U_{\infty}.\frac{d}{dx}\left(\frac{\theta^2}{\upsilon}\right) = -2\left[\frac{\theta^2}{\upsilon}.\frac{dU_{\infty}}{dx}\left(\frac{\delta^*}{\theta}+2\right) - \frac{\theta}{U_{\infty}}.\frac{\tau_0}{\mu}\right]$$

Then when $l = \frac{\theta\tau_0}{U_{\infty}\mu}$, $H = \frac{\delta^*}{\theta}$, $\lambda = \frac{\theta^2}{\upsilon}.\frac{dU_{\infty}}{dx}$

$$\therefore \quad U_{\infty}.\frac{d}{dx}\left(\frac{\lambda}{dU_{\infty}/dx}\right) = -2[\lambda(H+2) - l]$$

$$\text{or} \quad U_{\infty}.\frac{d}{dx}\left(\frac{\lambda}{dU_{\infty}/dx}\right) = 2\big(l - \lambda(H+2)\big) = f(\lambda)$$

$$\text{or} \quad U_{\infty}.\frac{d}{dx}\left(\frac{\lambda}{dU_{\infty}/dx}\right) = 0.45 - 6\,\lambda$$

$$U_{\infty}.\frac{d}{dx}\left(\frac{\lambda}{dU_{\infty}/dx}\right) + \frac{\lambda}{dU_{\infty}/dx}.\frac{dU_{\infty}}{dx} - \frac{\lambda}{dU_{\infty}/dx}.\frac{dU_{\infty}}{dx} = 0.48 - 6\,\lambda$$

$$\frac{d}{dx}\left[U_{\infty}.\frac{\lambda}{dU_{\infty}/dx}\right]$$

$$= 0.45 - 5\,\lambda$$

$$= 0.45 - 5\,\lambda.\frac{dU_{\infty}/dx}{dU_{\infty}/dx}.\frac{{U_{\infty}}^5}{{U_{\infty}}^5}$$

$$= 0.45 - \frac{\lambda}{dU_{\infty}/dx}.U_{\infty}.5{U_{\infty}}^4\frac{dU_{\infty}}{dx}.\frac{1}{{U_{\infty}}^5}$$

$$U_\infty{}^5 \frac{d}{dx}\left[U_\infty . \frac{\lambda}{dU_\infty/dx}\right] = 0.45U_\infty{}^5 - \frac{\lambda}{dU_\infty/dx} . U_\infty . \frac{dU_\infty{}^5}{dx}$$

$$U_\infty{}^5 \frac{d}{dx}\left[U_\infty . \frac{\lambda}{dU_\infty/dx}\right] = 0.45U_\infty{}^5 - \left(U_\infty . \frac{\lambda}{dU_\infty/dx}\right)\frac{dU_\infty{}^5}{dx}$$

or $\frac{d}{dx}\left[U_\infty{}^5 . \left(U_\infty . \frac{\lambda}{dU_\infty/dx}\right)\right] = 0.45U_\infty{}^5$

integrate both side

$$U_\infty{}^5 . \left(U_\infty . \frac{\lambda}{dU_\infty/dx}\right) = \int_0^x 0.45U_\infty{}^5 \, dx$$

$$\text{or } \lambda = \frac{0.45}{U_\infty{}^6} . \frac{dU_\infty}{dx} \int_0^x U_\infty{}^5 \, dx$$

b) To find the velocity profile

$U_0 W b = U(w + 2 \times \tan \emptyset)b$

or $U_{(x)} = U_0 \left(1 + \frac{x}{a}\right)^{-1}$

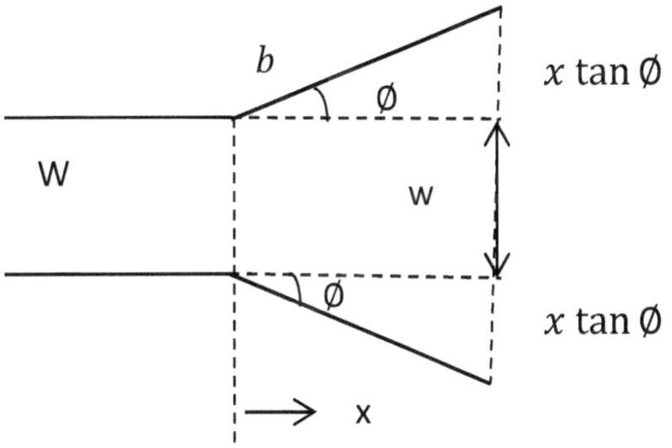

where $a = \frac{W}{2\tan \emptyset}$

$$\therefore \ \lambda = \frac{0.45}{U^6}\frac{\partial U}{\partial x} \int U^5 \, dx$$

Since $U = U_0 \left(1 + \frac{x}{a}\right)^{-1}$

$$\therefore \quad \frac{\partial U}{\partial x} = U_0\left(1+\frac{x}{a}\right)^{-2} \times -1 \times +\frac{1}{a}$$

$$= \frac{-U_0}{a}\left(1+\frac{x}{a}\right)^{-2}$$

$$\therefore \lambda = \frac{0.45}{U_0^{\,6}\left(1+\frac{x}{a}\right)^{-6}} \times \frac{-U_0}{a}\left(1+\frac{x}{a}\right)^{-2}\left[U_0^{\,5}\int_0^x \left(1+\frac{x}{a}\right)^{-5} dx\right]$$

$$\lambda = \frac{0.45}{U_0^{\,6}\left(1+\frac{x}{a}\right)^{-6}} \cdot \frac{-U_0}{a}\left(1+\frac{x}{a}\right)^{-2}\left[U_0^{\,5} \times \left(1+\frac{x}{a}\right)^{-4} \times \frac{1}{-4} \times \frac{a}{1}\right]_0^x$$

$$= \frac{0.45}{\left(1+\frac{x}{a}\right)^{-6}}\left(1+\frac{x}{a}\right)^{-2} \cdot \frac{1}{4}\left[\left(1+\frac{x}{a}\right)^{-4} - 1\right]$$

$$\lambda = \frac{0.45}{4}\left[1-\left(1+\frac{x}{a}\right)^{4}\right]$$

At the separation point $\lambda = -0.09$

$$\therefore \; -0.09 = \frac{0.45}{4}\left[1-\left(1+\frac{x}{a}\right)^{4}\right]$$

$$-0.8 = 1-\left(1+\frac{x}{a}\right)^{4}$$

$$-1.8 = -\left(1+\frac{x}{a}\right)^{4}$$

$$1.158 = 1+\frac{x}{a}$$

$$\therefore \; \left(\frac{x}{a}\right) = 0.1583 = \frac{L(2\tan \emptyset)}{W} \quad \text{at} \quad x = L$$

d) Since $L = 2\,W$

e)

$$\therefore \ \tan \emptyset = \frac{0.1583}{4}$$

$$\therefore \ \emptyset = 2.27°$$

Q.4). A long circular cylinder which is initially at rest and immersed in a stagnant fluid, is suddenly starts to rotates at constant speed U_∞. Due to viscosity, the fluid near the cylinder moves with it as a boundary layer with thickness δ. Assume the velocity profile to be linear. Find the boundary layer thickness and the stress at the wall.

Note: for unsteady flow $\tau_w = \rho \frac{\partial}{\partial t} \int_0^\delta (U_\infty - u)\, dy$

Solution:

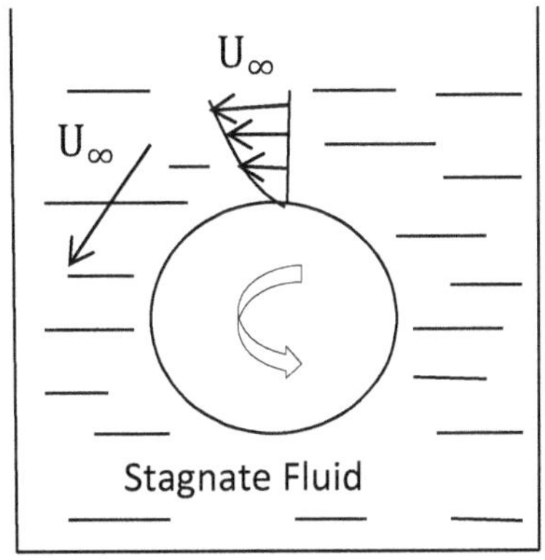

For Unsteady problem $\frac{\partial}{\partial t}$ exists

Since $\frac{\partial u}{\partial x} = 0$ same profile around the cylinder

constant U_∞, $\frac{\partial U_\infty}{\partial x} = 0$

and $\frac{\tau_0}{\rho} = \frac{\partial}{\partial t} \int_0^\delta (U_\infty - u)\, dy$

linear profile $\frac{u}{U_\infty} = \frac{y}{\delta}$

Now $\tau_0 = \mu \left. \frac{du}{dy} \right|_{y=0} = \mu \frac{U_\infty}{\delta}$ **(1)**

$$\tau_0 = \rho\, U_\infty \frac{\partial}{\partial t} \int_0^\delta \left(1 - \frac{u}{U_\infty}\right) dy$$

$$= \rho\, U_\infty \frac{\partial}{\partial t} \int_0^\delta \left(1 - \frac{y}{\delta}\right) dy$$

$$= \rho\, U_\infty \frac{\partial}{\partial t} \left[y - \frac{y^2}{2\delta}\right]_0^\delta = \rho\, U_\infty \frac{1}{2} \frac{\partial}{\partial t} (\delta)$$

or $\tau_0 = \frac{1}{2}\, \rho\, U_\infty \frac{\partial \delta}{\partial t}$ **(2)**

from 1,2

$$\mu \frac{U_\infty}{\delta} = \frac{1}{2}\, \rho\, U_\infty \frac{\partial \delta}{\partial t}$$

$$\int 2\, \upsilon\, dt = \int \delta\, d\delta$$

$$2\, \upsilon\, t = \frac{\delta^2}{2} \qquad , \delta = 2\sqrt{\upsilon\, t} = 2\sqrt{\frac{\mu}{\rho} t}$$

Then

$$\tau_0 = \mu \frac{U_\infty}{\delta} = \mu \frac{U_\infty}{2\sqrt{\frac{\mu}{\rho} t}} = \frac{1}{2} \sqrt{\frac{\mu {U_\infty}^2}{\frac{\mu}{\rho} t}}$$

$$\tau_0 = \frac{1}{2} \sqrt{\frac{\rho \mu}{t}}\, U_\infty$$

Q.5). A smooth flat plate 3m wide and 30m long is moving through still wake at 20^0 C with a speed of 6 m/s. Determined the drag on one side of the plate and the drag on first 3 m of the plate. $\rho = 998\ kg/m^3, \upsilon = 1.007 \times 10^{-6}\ m^2/s$.

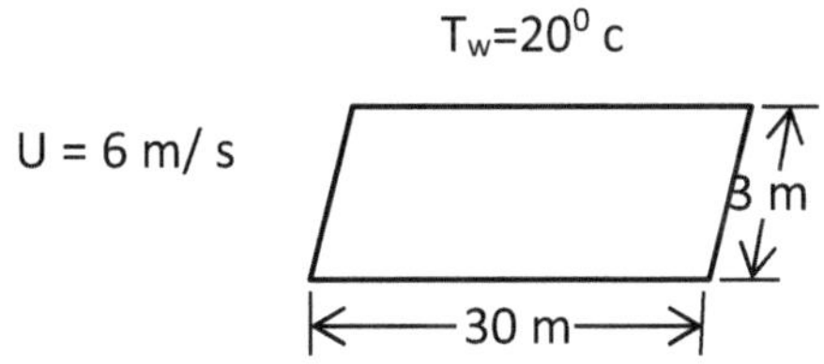

Solution:

$D = ?$

a) First we must test the flow.

$$Re = \frac{\rho UL}{\rho} = \frac{UL}{\upsilon} = \frac{6 \times 30}{1.007 \times 10^{-6}} = 1.78 \times 10^8 \quad \text{turbulat flow}$$

b) The Drag D $\Rightarrow$ for $10^6 < Re < 10^9$

$$C_D = \frac{0.455}{[\log(1.78 \times 10^8)]^{2.58}} = 0.00196$$

$$\therefore \ \mathrm{D} = c_D \times \frac{1}{2}\rho U^2 A = 0.00196 \times 998 \times (6)^2 \times (30 \times 3)$$

$\mathrm{D} = 3.071\ KN.$

c) For the first $L = 3 \Rightarrow Re = \frac{6\times3}{1.007\times10^{-6}} = 1.78 \times 10^7$ Turbulat flow

$$C_D = \frac{0.455}{(\log 1.78 \times 10^4)^{2.58}} = 0.0027$$

$$\mathrm{D} = 0.0027 \times \frac{1}{2} \times 998 \times (6)^2 \times (3 \times 3) = 0.437\ kN$$

Q.6). The velocity profile through the laminar boundary layer on a flat plate may be approximated by the expression.

$$\frac{u}{U} = A\eta - B\eta^3$$

Find (δ) at distance (1 m) from the leading edye. Where $A\&B$ is constants.
Solution:

Boundary condition :- $u = \bar{U}$ at $y = \delta \Rightarrow \eta = \frac{y}{\delta} = 1$

$\therefore \quad 1 = A - B$ **(1)**

2) $y = \delta \quad \frac{df(\eta)}{d\eta} = 0 . \quad i.e. \quad \frac{u}{U} = f(\eta)$

$\frac{df(\eta)}{d\eta} = A - 3B\eta^2 \quad \Rightarrow 0 = A - 3B$ or $A = 3B$ **(2)**

$\therefore$ From 1,2 $1 = 3B - B \quad \Rightarrow \quad B = \frac{1}{2} \ , \ A = \frac{3}{2}$

$$\therefore \quad \frac{u}{U} = \frac{3}{2}\eta - \frac{1}{2}\eta^3$$

$$\because \ \tau_w = \rho U^2 \frac{\partial}{\partial x} \int_0^\delta \frac{u}{U}\left(1 - \frac{u}{U}\right) dy$$

$$= \rho U^2 \frac{\partial \delta}{\partial x} \int_0^1 \left(\frac{3}{2}\eta - \frac{1}{2}\eta^3\right)\left(1 - \frac{3}{2}\eta + \frac{1}{2}\eta^3\right) d\eta$$

$$= \rho U^2 \frac{\partial \delta}{\partial x} \times 0.139$$

To find the B.L. thickness

$$\tau_w = \mu \left.\frac{du}{dy}\right|_{y=0}$$

$$u = U\left[\frac{3}{2}\eta - \frac{1}{2}\eta^3\right] = U\left[\frac{3}{2}\left(\frac{y}{\delta}\right) - \frac{1}{2}\left(\frac{y}{\delta}\right)^3\right]$$

$$\frac{du}{dy} = U\left[\frac{3}{2\delta} - \frac{3}{2}\frac{y^2}{\delta^2}\right] \Rightarrow \left.\frac{\partial u}{\partial y}\right|_{y=0} U\left[\frac{3}{2\delta} - 0\right]$$

$$\therefore\ \tau_w = 1.5\frac{U}{\delta}\mu$$

$$\therefore\ 1.5\frac{U}{\delta}\mu = 0.139\ \rho U^2\frac{\partial\delta}{\partial x}$$

$$\Rightarrow\ \delta = \frac{4.64\ x}{\sqrt{Re_x}}$$

Q.7).water flows at velocity of 7 m/s on a flat plate 30 m long and 4 m wide. Find the drag force at the trailing edge and at 3 m from the leading edge? *i.e.* $\mu = 0.001\ Pa.s$

Solution:

1- Test the type of flow:

$$Re = \frac{\rho UL}{\mu} = \frac{1000 \times 7 \times 30}{0.001} = 21 \times 10^7$$

2- The Drag coefficient C_D for $10^6 < Re < 10^9$

$$C_D = \frac{0.455}{(\log Re)^{2.56}} = \frac{0.455}{[\log(21 \times 10^7)]^{2.56}} = 2 \times 10^{-3}$$

3- The Drag force D:

$$D = C_D \times \frac{1}{2} \rho U^2 A$$

$$= 2 \times 10^{-3} \times \frac{1}{2} \times 1000 \times (7)^2 \times (30 \times 4) \times 2$$

$D = 11302\ N$

4- For the length $L = 3\ m$ and $w = 4\ m$

a) $Re = \frac{1000 \times 7 \times 3}{0.001} = 21000000 = 21 \times 10^6$

b) $C_D = \frac{0.455}{(\log 21 \times 10^6)^{2.56}} = 2.78 \times 10^{-3}$

c) $D = 2.78 \times 10^{-3} \times 0.5 \times 1000 \times (7)^2 \times (3 \times 4) \times 2$

$D = 1571.5\ N$

Q.8). An air flows on flat plate at the following

$$U = 120\ m/s, \rho = 1.225, \mu = 1.789 \times 10^{-5}, L = 5\ cm$$

Find:

a) Type of B.L.

b) Thickness at the end of the plate.

c) The drag force on the plate.

Solution:

a) $Re = \frac{\rho UL}{\mu} = \frac{1.225 \times 120 \times 5 \times 10^{-2}}{1.789 \times 10^{-5}} = 412921.34 = 4.1292134 \times 10^5$

$\therefore$ the flow is laminar boundary layer.

b) $\delta = \frac{4.64\, x}{\sqrt{Re}} = \frac{4.64 \times 5 \times 10^{-2}}{\sqrt{412921.34}} = 0.36\, mm.$

c) $D = C_D \times \frac{1}{2} \rho U^2 A$

$$C_D = 1.288/\sqrt{Rex} = 1.288/\sqrt{412921.34} = 2 \times 10^{-3}$$

$$\therefore D = 2 \times 10^{-3} \times \frac{1}{2} 1.225 \times (120)^2 \times (0.05 \times 1) \times 2$$

$$D = 1.77\, N$$

Q.9). An air flowing throughout a square duct for the following

$$U = 4\ m/s, \qquad \upsilon = 1.25 \times 10^{-6}\ m^2/s, \qquad \frac{u}{U} = \left(\frac{y}{\delta}\right)^{1/7}$$

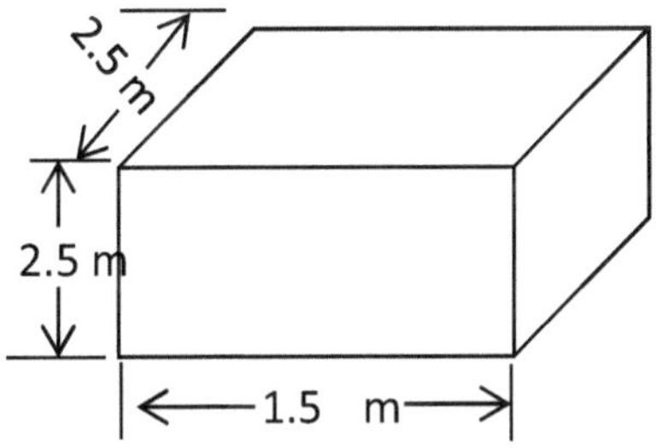

Assume turbulent flow B.L.?

Find:

δ^* & $\Delta P = ?$

Solution:

a) $\delta^* =$ for turbulent flow B. L. $= \frac{\delta}{8}$

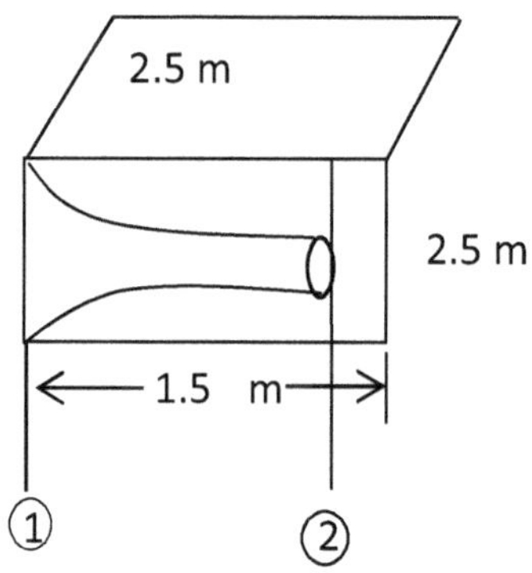

Now: $\delta = 0.37 \times X \times Re^{-\frac{1}{5}}$

$$\therefore\ \delta = 0.37 \times 1.5 \times \left(\frac{1.25 \times 10^{-6}}{4 \times 1.5}\right)^{-\frac{1}{5}}$$

$$\therefore\ \delta = 0.026\ m$$

$$\therefore\ \delta^* = \frac{\delta}{8} = 3.2 \times 10^{-3}\ m$$

b) To find the pressure between the suction 1 & 2 we apply B.E.

$$\frac{P_1}{\rho g} + \frac{{u_1}^2}{2g} + z_1 = \frac{P_2}{\rho g} + \frac{{u_2}^2}{2g} + z_2\ .$$

$$A_1 u_1 = A_2 u_2\ \Rightarrow\ (2.5)^2 \times 4 = (2.5 - 2 \times \delta^*)2.5 \times u_2$$

$$\Rightarrow\ u_2 = 4.16\ m/s$$

$$\therefore\ (P_1 - P_2) = \frac{\rho}{2}({u_2}^2 - {u_1}^2)$$

$$\Delta P = 41.2\ N/m^2$$

Q.10). Find L=?

$$T_a = 150^{\circ}, \qquad P_a = 1.013\ bar, \qquad U = 100\ Km/hr,$$

$$L = ?, \delta = 0.6\ cm \text{ for laminar B. L. } \upsilon = 1.8 \times 10^{-5}\ m^2/s$$

Solution:

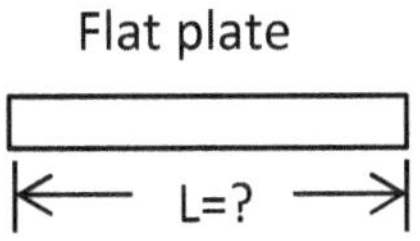

$$\text{Now } Re = \frac{\rho_a UL}{\mu}$$

$$= \frac{UL}{\upsilon} = \frac{100 \times \frac{1000}{3600} \times L}{1.8 \times 10^{-5}} = 15{,}43209.9\ L$$

$$\text{Now for laminar } B.L.\ \delta = \frac{4.64 \times L}{\sqrt{Rex}}$$

$$\therefore 0.6 \times 10^{-2} = \frac{4.64 \times L}{\sqrt{Rex}} =$$

$$0.6 \times 10^{-2} = 3.735 \times 10^{-3}\sqrt{L}$$

$$\Rightarrow L = 2.58\ m$$

Q.11). Find the drag coefficient at the transition zone?

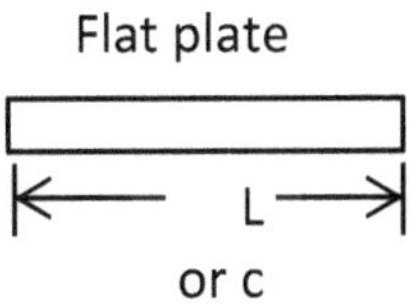

$C_D = ?$

$$\frac{u}{U} = \frac{3}{2}\bar{y} - \frac{1}{2}\bar{y}^3 \qquad \text{Laminar} \qquad \bar{y} = \frac{y}{\delta}$$

$$\frac{u}{U} = \bar{y}^{\frac{1}{7}} \qquad \text{Turbulent}$$

Solution:

* for Laminar flow : − $\frac{u}{U} = \frac{3}{2}\frac{y}{\delta} - \frac{1}{2}\left(\frac{y}{\delta}\right)^3$

So the B. L. thickness $\delta_L = \frac{4.64\, x}{\sqrt{Rex}}$

and the momentum thickness $\theta_L = 0.139\, \delta_L$

* for Turbulent flow : − $\frac{u}{U} = \left(\frac{y}{\delta}\right)^{\frac{1}{7}}$

So the B. L. thickness $\delta_T = 0.37\, X_T\, Rex^{-\frac{1}{5}}$

and the momentum thickness $\theta_T = 0.0973\, \delta_T$

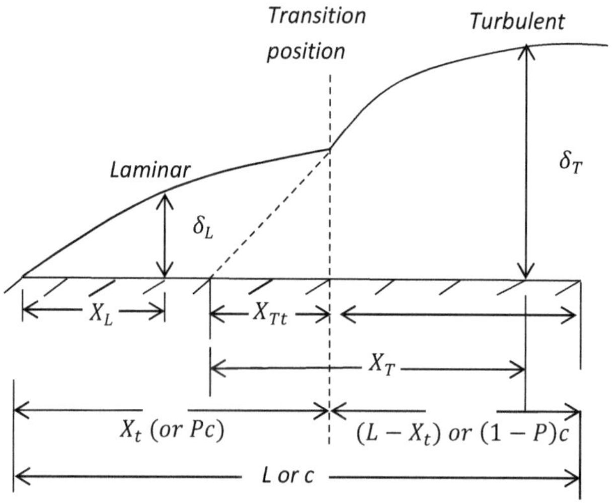

At Transition $\theta_L = \theta_T$

$$0.0973\ \delta_T = 0.139\ \delta_L$$

$$\therefore\ \frac{\delta_T}{\delta_L} = \frac{0.139}{0.0973} = 1.432$$

Now to find δ_T & δ_L at the distance Pc from the leading edge.

$$\therefore\ \delta_L = 4.64\ \times X_t \left(\frac{\upsilon}{UX_t}\right)^{\frac{1}{2}} = 4.64\left(\frac{\upsilon}{U}\right)^{\frac{1}{2}}(Pc)^{\frac{1}{2}}$$

$$\delta_T = 0.37\ \times X_{Tt}\left(\frac{\upsilon}{UX_t}\right)^{\frac{1}{5}} = 0.37(X_{Tt})^{\frac{4}{5}}\left(\frac{\upsilon}{U}\right)^{\frac{1}{5}}$$

$$\therefore\ 0.37(X_{Tt})^{\frac{4}{5}}\left(\frac{\upsilon}{U}\right)^{\frac{1}{5}} = 1.432 \times 4.64 \times \left(\frac{\upsilon}{U}\right)^{\frac{1}{2}}(P_c)^{\frac{1}{2}}$$

$$\therefore\ X_{Tt} = 37(P_c)^{\frac{5}{8}}\left(\frac{\upsilon}{u}\right)^{\frac{3}{8}}\quad \text{from the starting of transition zone.}$$

$$\therefore\ C_D = 0.072Re^{-\frac{1}{5}}$$

$$= 0.072\left(\frac{\upsilon}{UX_{Tt}}\right)^{-\frac{1}{5}}$$

By substituting the distance of transition zone X_{Tt} to find the drag coefficient we should multiply by the ratio

$$\frac{X_{Tt} + (1 - P)c}{c}$$

Then

$$C_D = \frac{D}{\frac{1}{2}\rho V^2 A}\quad \text{or}\quad C_D = \frac{0.072}{Re^{\frac{1}{5}}} \times \frac{X_{Tt} + (1 - P)c}{c}$$

where $Re = f\left(\frac{X_{Tt} + (1-P)c}{c}\right)$

$$C_D = \frac{0.072}{\left(\frac{U}{\upsilon}\right)^{\frac{1}{5}} c} \times (X_{Tt} + (1-P)c)^{\frac{4}{5}} \times \frac{\left(\frac{U}{\upsilon}\right)^{\frac{4}{5}}}{\left(\frac{U}{\upsilon}\right)^{\frac{4}{5}}} = \frac{0.072(X_{Tt} + (1-P)c)^{\frac{4}{5}}}{\left(\frac{U}{\upsilon}\right) c}$$

$$C_D = \frac{0.072(X_{Tt} + (1-P)c)^{\frac{4}{5}}}{\left(\frac{Uc}{\upsilon}\right)} = \frac{0.072\left[\frac{U}{\upsilon} X_{Tt} + (1-P)c\right]^{\frac{4}{5}}}{Re}$$

$$X_{Tt} = 37(Pc)^{\frac{4}{5}} \left(\frac{\upsilon}{U}\right)^{\frac{3}{8}}$$

Q.12) For the previous equation find P when the drag force at the transition zone is equal. $D_{Tt} = D_t - 4.5$

$$U = 45\ m/s\ , L = 1.8\ m = c\ , D_{Tt} = D_t - 4.5$$

Solution:

$$\therefore \mathrm{D} = C_D \frac{1}{2} \rho U^2 A$$

Now for the leading edge *i.e.* for all $c = 1.8\ m$

$$Re = \frac{UL}{\upsilon} = \frac{45 \times 1.8}{1.46 \times 10^{-5}} = 55.5 \times 10^5$$

$$\therefore C_D = \frac{0.072}{Re^{\frac{1}{5}}} = \frac{0.072}{(55.5 \times 10^5)^{\frac{1}{5}}} = 3.22 \times 10^{-3}$$

$$\therefore D_t = 3.22 \times 10^{-3} \times \frac{1}{2} \times 1.225 \times (45)^2 \times (1.8) = 7.2\ N$$

Now, with Transition $D_{Tt} = 7.2 - 4.5 = 2.7\ N$

$$2.7 = C_D \frac{1}{2} \rho U^2 A \quad \text{or} \quad C_D = \frac{2.7}{\frac{1}{2} \times 1.225 \times (45)^2 \times (1.8)} = 1.21 \times 10^{-3}$$

For transition zone $C_D = \frac{0.072}{Re_c} \left[3.7P^{\frac{5}{8}} \times Re_c^{\frac{5}{8}} + (1-P)Re_c \right]^{\frac{4}{5}}$

Now $C_D = \frac{0.072}{Re_c} \left[3.7P^{\frac{5}{8}} \times 16412 + Re_c(1-P) \right]^{\frac{4}{5}}$

$$1.21 \times 10^{-3} = \frac{0.072}{55.5 \times 10^5} \left[37P^{\frac{5}{8}} \times 16412 + Re_c(1-P) \right]^{\frac{4}{5}}$$

$P \approx 1$ Trie and error

Q..13). For flow on a flat plate find the boundary Layer thickness and the drag force?

$L = 0.6\ m, \quad U = 45\ m/s, \quad \frac{u}{U} = 2\frac{y}{\delta} - \left(\frac{y}{\delta}\right)^2$

Find: $(i)\ \delta \quad (ii)\ D$

Solution:

$$\frac{u}{U} = 2\left(\frac{y}{\delta}\right) - \left(\frac{y}{\delta}\right)^2$$

Let $\left(\frac{y}{\delta}\right) = \eta \quad \Rightarrow dy = \delta\, d\eta$

Since $\delta = \int_0^{\delta} \left(1 - \frac{u}{U}\right) dy$

$$\therefore \quad \delta = \int_0^1 (1 - 2\eta + \eta^2)\, dy$$

and $$\tau_w = \rho U^2 \frac{\partial}{\partial x} \int_0^\delta \frac{u}{U}\left(1 - \frac{u}{U}\right) dy$$

$$= \rho U^2 \frac{\partial \delta}{\partial x} \int_0^1 (2\eta + \eta^2)(1 - 2\eta + \eta^2)\, d\eta$$

$$= \rho U^2 \frac{\partial \delta}{\partial x} \int_0^1 [2\eta - 4\eta^2 + 2\eta^3 - \eta^2 + 2\eta^3 - \eta^4]\, d\eta$$

$$= \rho U^2 \frac{\partial \delta}{\partial x} \int_0^1 [2\eta - 5\eta^2 + 4\eta^3 - \eta^4]\, d\eta$$

$$= \rho U^2 \frac{\partial \delta}{\partial x} \left[\eta^2 - \frac{5}{3}\eta^3 + \eta^4 - \frac{1}{5}\eta^5\right]_0^1$$

$$\tau_w = 0.133\, \rho U^2 \frac{\partial \delta}{\partial x}$$

$$\because \quad \tau_w = \mu \left.\frac{\partial u}{\partial y}\right|_{y=0} \quad \Rightarrow \quad \frac{\partial u}{\partial y} = U\left[\frac{2}{\delta} - \frac{2y}{\delta^2}\right]$$

$$\frac{\mu 2U}{\delta} = 0.133\, \rho U^2 \frac{\partial \delta}{\partial x} \quad \Rightarrow \quad \frac{2 \times 7.5\, \upsilon}{U} \int dx = \int \delta\, \partial\delta$$

$$= 15 \frac{\upsilon}{U} x = \frac{\delta^2}{2}$$

$$\Rightarrow \quad \delta = \sqrt{30}\left(\frac{\upsilon}{U} x\right)^{\frac{1}{2}}$$

$$\delta = \frac{5.47\,x}{\sqrt{Rex}} = \frac{5.47 \times 0.6}{\left(\frac{45 \times 0.6}{1.46 \times 10^{-5}}\right)^{\frac{1}{2}}} = 2.41 \times 10^{-3}\,m$$

b)

$$D = \int_0^L \tau_w\ dx$$

$$\therefore D = \int_0^L \rho U^2 \frac{\partial}{\partial x}\left[\int_0^\delta \frac{u}{U}\left(1 - \frac{u}{U}\right) dy\right] dx$$

$$\Rightarrow\ \ D = f(\tau_w)$$

Now: $\tau_w = 0.133\,\rho U^2 \frac{\partial \delta}{\partial x}$

$$\frac{\partial \delta}{\partial x} = 5.47\left(\frac{\upsilon}{U}\right)^{\frac{1}{2}} \times \frac{1}{2} x^{\frac{1}{2}} = 2.738\,Re^{-\frac{1}{2}}$$

$$\therefore\ \ D = \int_0^L 0.133\,\rho U^2 \times 2.738\left(\frac{\upsilon}{U}\right)^{\frac{1}{2}} x^{-\frac{1}{2}}\,dx$$

$$D = 0.364\,\rho U^2\left(\frac{\upsilon}{U}\right)^{\frac{1}{2}} \times \left.\frac{x^{\frac{1}{2}}}{\frac{1}{2}}\right|_0^L$$

$$D = 0.728\,\rho U^2\left(\frac{\upsilon L}{U}\right)^{\frac{1}{2}}$$

$$D = 0.728 \times 1.225 \times (45)^2\left(\frac{1.46 \times 10^{-5} \times 0.6}{45}\right)^{\frac{1}{2}} = 0.8\,N$$

Q.14). Using the integral momentum analysis of the flat plate, the assumed velocity profile:

$$\frac{u}{U} = \frac{3}{2}\left(\frac{y}{\delta}\right) - \frac{1}{2}\left(\frac{y}{\delta}\right)^3$$

Where δ is the B.L. thickness, compute

$$a)\ \frac{\theta}{x}\sqrt{Re_x} \quad b)\frac{\delta^*}{x}\sqrt{Re_x} \quad c)\frac{\delta}{x}\sqrt{Re_x} \quad d)C_f\sqrt{Re_x} \quad e)C_D\sqrt{Re_x}$$

Solution:

$$\frac{u}{U} = \frac{3}{2}\left(\frac{y}{\delta}\right) - \frac{1}{2}\left(\frac{y}{\delta}\right)^3$$

$$= \frac{3}{2}\eta - \frac{1}{2}\eta^3 \quad = f(\eta)$$

$$B = \frac{df(\eta)}{d\eta} = \frac{3}{2} - \frac{3}{2}\eta^2\Big|_{\eta=0} = \frac{3}{2}$$

$$A = \int_0^1 f(\eta)(1 - f(\eta))\, d\eta = \int_0^1 \left(\frac{3}{2}\eta - \frac{1}{2}\eta^3\right)\left(1 - \frac{3}{2}\eta - \frac{1}{2}\eta^3\right) d\eta$$

$$= \int_0^1 \left(\frac{3}{2}\eta - \frac{1}{2}\eta^3 - \frac{9}{4}\eta^2 + \frac{3}{4}\eta^4 + \frac{3}{4}\eta^4 - \frac{1}{4}\eta^6\right) d\eta$$

$$= \left[\frac{3}{2}\eta^2 - \frac{1}{8}\eta^4 - \frac{9}{12}\eta^3 + \frac{3}{20}\eta^5 + \frac{3}{20}\eta^5 - \frac{1}{28}\eta^7\right]_0^1$$

$$= \left[\frac{3}{2} - \frac{1}{8} - \frac{9}{12} + \frac{3}{20} + \frac{3}{20} - \frac{1}{28}\right]_0^1 = 0.14$$

$A = 0.14$

$$\delta = \sqrt{\frac{2B}{A} \cdot \frac{x}{Re_x}} = \sqrt{\frac{2 \times 1.5}{0.14}} \cdot \frac{x}{\sqrt{Re_x}} = 4.63 \frac{x}{\sqrt{Re_x}}$$

or $\frac{\delta}{x} \cdot \sqrt{Re_x} = 4.63$

$$c)\ \tau_0 = \rho U^2 \sqrt{\frac{AB}{2Re_x}} = \rho U^2 \sqrt{\frac{0.14 \times 1.5}{2}} \cdot \frac{1}{\sqrt{Re_x}}$$

$$\tau_0 = \frac{0.324 \rho U^2}{\sqrt{Re_x}}$$

$$\therefore C_f = \frac{\tau_0}{\frac{1}{2}\rho U^2} = \frac{0.648}{\sqrt{Re_x}}$$

$$d)\ C_f \sqrt{Re_x} = 0.648$$

$$F_D = \sqrt{2AB\mu\rho U^3 L} = \sqrt{2 \times 0.14 \times 1.5\ \mu\rho U^3 L}$$

$$F_D = 0.648 \sqrt{\mu\rho U^3 L}$$

$$C_D = \frac{F_D}{\frac{1}{2}\rho U^2 L} = 2 \times 0.648 \sqrt{\frac{\mu\rho U^3 L}{\rho^2 U^4 L^2}}$$

$$C_D = 1.296 \sqrt{\frac{\mu}{\rho U L}}$$

$$e)\ C_D \sqrt{Re_L} = 1.296$$

$$a)\ \theta = A\delta \quad \Rightarrow \frac{\theta}{x}\sqrt{Re_x} = A \times 4.63 = 0.14 \times 4.63 = 0.649$$

$$b)\ \delta^* = \delta\int_0^1 (1 - f(\eta))\, d\eta = \delta\int_0^1 \left(1 - \frac{3}{2}\eta + \frac{1}{2}\eta^3\right) d\eta$$

$$\delta^* = \delta\left[\eta - \frac{3}{4}\eta^2 + \frac{1}{8}\eta^4\right]_0^1$$

$$= \delta\left[1 - \frac{3}{4} + \frac{1}{8}\right] = 0.375\ \delta$$

$$b)\ \delta^* = 0.375\ \times \frac{x \times 4.63}{\sqrt{Re_x}}$$

$$\delta^* = 1.736\ \times \frac{x}{\sqrt{Re_x}}$$

$$\therefore \frac{\delta^*}{x}\sqrt{Re_x} = 1.736$$

Q.15). Air at 20 ^{0}C and 100 kPa flow along rectangular smooth flat with 0.3 m width and 1 m length. Evaluate shear stress at the mid of the plate when $Re_x = 10^6$, the drag coefficient and drag force for the whole plate if the flow is completely turbulent over the plate.

$$C_f = \frac{0.059}{Re_x^{\ 0.2}}, v_{air} = 15\ mm^2/sec\,, R = 287\ J/kg\,.K$$

Solution:

$$\rho = \frac{P}{RT} = \frac{100 \times 10^3}{287 \times 293} = 1.189\ kg/m^3$$

$$Re = \frac{U_\infty x}{v} = 10^6 = \frac{U_\infty \times 0.5}{15 \times 10^{-6}} \quad \Rightarrow \quad U_\infty = 30\ m/s$$

$$\therefore\ C_{f\ at\ mid\ point} = \frac{0.059}{\left(\frac{U_\infty x}{v}\right)^{1/5}} \qquad x = 0.5$$

$$\therefore \; C_f = 3.723 \times 10^{-3}$$

$$\tau_{w\,at\,mid\,point} = \frac{1}{2}\rho U^2 C_f = 1.99 \; N/m^2$$

$$Re_L = \frac{U_\infty L}{\upsilon} = \frac{30 \times 1}{15 \times 10^{-6}} = 2 \times 10^6 < 10^7$$

$$\therefore \; C_D = \frac{0.072}{(Re_L)^{1/5}} = 3.9548 \times 10^{-3}$$

$$\therefore \; F_D = \frac{1}{2}\rho {U_\infty}^2 C_D \times L \times B \times 2$$

$$= 1.269 \; N \,.$$

Q.16). the integral B.L eq. Written as follows:

$$\frac{C_f}{2} = \frac{\partial \theta}{\partial x} + \frac{\theta}{U}\frac{\partial U}{\partial x}(2 + H)$$

Where : $H = \frac{\delta^*}{\theta}$ Shape factor

$\lambda = \frac{\theta^2}{U}\frac{\partial U}{\partial x}$ Pressure gradient factor

$S_{(\lambda)} = \frac{\theta \tau_w}{\mu U}$ Shear stress parameter

Using Thwaites straight-line correlation

$f_{(\lambda)} = 0.45 - 6\lambda$ and $f_{(\lambda)} = 2\left[S_{(\lambda)} - \lambda(2 + H)\right]$

a) Show that $\lambda = \frac{0.45}{U^6}\frac{\partial U}{\partial x}\int_0^x U^5 . dx$

b) If the proposed external velocity distribution $U = U_\infty\left(1 - \frac{x}{L}\right)$

Find λ at $\frac{x}{L} = 0.1$ and separation point $\frac{x_{sep}}{L}$ at $\lambda_{sep} = -0.09$

Solution:

$$\frac{\tau_0}{\rho U^2} = \frac{\partial\theta}{\partial x} + \frac{\theta}{U}\frac{\partial U}{\partial x}(2 + H)$$

$$\frac{\partial\theta}{\partial x} = \frac{\tau_0}{\rho U^2} - +\frac{\theta}{U}\frac{\partial U}{\partial x}(2 + H)$$

$$U\frac{\partial\theta}{\partial x} = \frac{\tau_0}{\rho U} - \theta\frac{\partial U}{\partial x}(2 + H) \qquad \times U$$

$$U\theta\frac{\partial\theta}{\partial x} = \frac{\theta\tau_0}{\rho U} - \theta^2\frac{\partial U}{\partial x}(2 + H) \qquad \times \theta$$

$$U\frac{\partial}{\partial x} = 2\left[\frac{\theta\tau_0}{\mu U} - \frac{\theta^2}{v}\frac{\partial U}{\partial x}(2 + H)\right] \qquad \times\frac{1}{v}$$

Since $l = \frac{\theta\tau_0}{\mu U}, \quad H = \frac{\delta^*}{\theta}, \quad \lambda = \frac{\theta^2}{v}\frac{\partial U}{\partial x}$

$$\therefore \quad U\frac{\partial}{\partial x}\left(\frac{\lambda}{\partial U/\partial x}\right) = 2[l - \lambda(2 + H)] = f(\lambda)$$

or $U\frac{\partial}{\partial x}\left(\frac{\lambda}{\partial U/\partial x}\right) = 0.45 - 6\lambda$

Since $U\frac{\partial}{\partial x}\left(\frac{\lambda}{\partial U/\partial x}\right) + \frac{\lambda}{\partial U/\partial x}\frac{\partial U}{\partial x} - \frac{\lambda}{\partial U/\partial x}\frac{\partial U}{\partial x} = 0.45 - 6\lambda$

Adding and subtracted to get

$$\therefore \frac{\partial}{\partial x}\left[U.\frac{\lambda}{\partial U/\partial x}\right] = 0.45 - 5\lambda\ \frac{\partial U/\partial x}{\partial U/\partial x}\cdot\frac{U^5}{U^5}$$

$$= 0.45 - \frac{\lambda}{\partial U/\partial x}.U.5U^4\frac{\partial U}{\partial x}.\frac{1}{U^5}$$

or $$U^5\frac{\partial}{\partial x}\left[U.\frac{\lambda}{\partial U/\partial x}\right] = 0.45U^5 - \frac{\lambda}{\partial U/\partial x}.U\frac{\partial U^5}{\partial x}$$

or $$\frac{\partial}{\partial x}\left[U^5.\left(U\frac{\lambda}{\partial U/\partial x}\right)\right] = 0.45U^5$$

integrate both side $$U^5.\left(U\frac{\lambda}{\partial U/\partial x}\right) = \int_0^x 0.45U^5\,dx$$

$$\therefore\ \lambda = \frac{0.45}{U^6}.\frac{\partial U}{\partial x}\int_0^x U^5\,dx$$

b)

Since $U = U_\infty\left(1 - \frac{x}{L}\right)$

$$\frac{\partial U}{\partial x} = -\frac{U_\infty}{L}$$

and $$\int_0^x {U_\infty}^5\left(1 - \frac{x}{L}\right)dx = {U_\infty}^5\left[\left(1 - \frac{x}{L}\right)^6\right]_0^x - \frac{L}{6}$$

then

$$\lambda = \frac{0.45}{{U_\infty}^6\left(1-\frac{x}{L}\right)^6} \times -\frac{U_\infty}{L} \times \frac{{U_\infty}^5}{6}\,L\left[\left(1 - \frac{x}{L}\right)^{-6} - 1\right]$$

$$\lambda = 0.075\left[1 - \left(1 - \frac{x}{L}\right)^{-6}\right]$$

(1) find $\lambda \;\; at \; \frac{x}{L} = 0.1$

$$\lambda = 0.075\left[1 - \left(1 - \frac{x}{L}\right)^{-6}\right]$$

$$\lambda = 0.075[1 - (1 - 0.1)^{-6}] = -0.0661$$

(2) at $\lambda = -0.09 = 0.075\left[1 - \left(1 - \frac{x_{sep}}{L}\right)^{-6}\right]$

$$-1.2 = 1 - \left(1 - \frac{x_{sep}}{L}\right)^{-6}$$

$$-2.2 = -\left(1 - \frac{x_{sep}}{L}\right)^{-6}$$

$$(2.2)^{-\frac{1}{6}} = 1 - \frac{x_{sep}}{L}$$

$$\therefore \;\; \frac{x_{sep}}{L} = 1 - (2.2)^{-\frac{1}{6}} = 0.123$$

Q.17). For the fourth order polynomial of B.L. velocity profile of flow is:

$$u = a + by + cy^2 + dy^3 + ey^4$$

Find, the displacement thickness as a function of Re_x.

Solution:

Fourth order polynomial is

$$u = a + by + cy^2 + dy^3 + ey^4$$

B. C. 1	at $y = 0$	$u = 0$	$\therefore$	$a = 0$
B. C. 2	at $y = \delta$	$\frac{du}{dy} = 0$	$\therefore$	$b = -3d\delta^2 - 4e\delta^3$

For flat plate only $\dfrac{\partial P}{\partial y} = 0$

B.C. 3 at $y = 0$ $\dfrac{\partial^2 u}{dy^2} = 0$ $\therefore\ c = 0$

B.C. 4 at $y = \delta$ $\dfrac{\partial^2 u}{dy^2} = 0$ $\therefore\ d = -2e\delta$

B.C. 5 at $y = \delta$ $u = U_\infty$ $\therefore U_\infty = b\delta + d\delta^3 + e\delta^4$

from the above relations gives

$$e = \frac{U_\infty}{\delta^4}, \qquad b = 2\frac{U_\infty}{\delta}, \qquad d = -2\frac{U_\infty}{\delta^3}$$

Then fourth order polynomial velocity profile

$$\frac{u}{U_\infty} = 2\frac{y}{\delta} - 2\left(\frac{y}{\delta}\right)^3 + \left(\frac{y}{\delta}\right)^4$$

Now $\dfrac{u}{U_\infty}\left(1 - \dfrac{u}{U_\infty}\right) = (2\eta - 2\eta^3 + \eta^4)(1 - 2\eta + 2\eta^3 - \eta^4)$

For $f(\eta) = \dfrac{u}{U_\infty}$ $\eta = \dfrac{y}{\delta}$ $dy = \delta\, d\eta$

Then

$$\frac{u}{U_\infty}\left(1 - \frac{u}{U_\infty}\right) = 2\eta - 4\eta^2 - 2\eta^3 + 9\eta^4 - 4\eta^5 + 4\eta^6 + 4\eta^7 - \eta^8$$

Since $\tau_0 = \rho U^2 \dfrac{\partial \delta}{\partial x}\, \alpha$ **(1)**

$$\alpha = \int_0^1 f(\eta)\big(1 - f(\eta)\big)\, d\eta$$

$$= \int_0^1 (2\eta - 2\eta^3 + \eta^4)(1 - 2\eta + 2\eta^3 - \eta^4)\, d\eta$$

$$= \int_0^1 (2\eta - 4\eta^2 + 4\eta^4 - 2\eta^5 - 2\eta^3 + 4\eta^4 - 4\eta^6 + 2\eta^7 + \eta^4 - 2\eta^5 + 2\eta^7 - \eta^8)\, d\eta$$

$$= \left[\frac{2\eta^2}{2} - \frac{4\eta^3}{3} + \frac{4\eta^5}{5} - \frac{2\eta^6}{6} - \frac{2\eta^4}{4} + \frac{4\eta^5}{5} + \frac{2\eta^8}{8} - \frac{4\eta^7}{7} + \frac{\eta^5}{5} - \frac{2\eta^6}{6} + \frac{2\eta^8}{8} - \frac{\eta^9}{9}\right]_0^1$$

$$= 1' - \frac{4'}{3} + \frac{4'}{5} - \frac{1'}{3} - \frac{1'}{2} + \frac{4'}{5} + \frac{1'}{4} + \frac{1'}{5} - \frac{1'}{3} + \frac{1'}{4} - \frac{4}{7} - \frac{1}{9}\Big]$$

$$= 1 - 2 + \frac{9}{5} - \frac{1}{2} + \frac{1}{2} - \frac{4}{7} - \frac{1}{9}$$

$$= 0.1175$$

$$\tau = \mu \frac{du}{dy} \tag{2}$$

And

$$u = U_\infty \left[2\frac{y}{\delta} - 2\left(\frac{y}{\delta}\right)^3 + \left(\frac{y}{\delta}\right)^4\right]$$

$$\frac{du}{dy} = U_\infty \left[\frac{2}{\delta} - 3 \times 2\frac{y^2}{\delta^3} + 4\frac{y^3}{\delta^4}\right]_{y=0}$$

$$= \frac{2U_\infty}{\delta}$$

$$\tau = \frac{2U_\infty \mu}{\delta} \tag{3}$$

$$0.1175\ \rho U_\infty^2 \frac{d\delta}{dx} = \frac{2U_\infty \mu}{\delta}$$

$$\delta\ d\delta = \frac{2}{0.1175} \cdot \frac{1}{\rho U}\ dx$$

$$\delta^2 = \frac{2}{0.1175} \cdot \frac{\mu}{\rho U}\ x$$

$$\delta^2 = \frac{\mu 34}{\rho U}\ x \qquad\qquad \delta = 5.83\sqrt{\frac{\mu x}{\rho U}}$$

or $\delta = 5.83x\sqrt{\dfrac{\mu}{\rho U x}}$

or $\delta = 5.83x\sqrt{\dfrac{\mu}{\rho U x}}$

or $\dfrac{\delta}{x} = \dfrac{5.83}{\sqrt{Re_x}}$

$$\delta^* = \delta \int_0^1 \big(1 - f(\eta)\big)\, d\eta$$

$$= \delta \int_0^1 (1 - 2\eta + 2\eta^3 - \eta^4)\, d\eta$$

$$= \delta \left[\eta - \frac{2\eta^2}{2} + \frac{2\eta^4}{4} - \frac{\eta^5}{5}\right]_0^1$$

$$= \delta\left[1 - 1 + \frac{1}{2} - \frac{1}{5}\right] = \frac{3}{10}\delta$$

$$\therefore \delta^* = \frac{3}{10} \times \frac{5.83\,x}{\sqrt{Re_x}} = \frac{1.755\,x}{\sqrt{Re_x}}$$

$$\theta = \delta \int_0^1 f(\eta)\big(1 - f(\eta)\big)\,d\eta$$

$$= 0.1175 \times \frac{5.83\,x}{\sqrt{Re_x}}$$

$$\theta = \frac{0.687\,x}{\sqrt{Re_x}}$$

Q.18). A block of 25 m long and 10 m wide submerged to a depth of 1.5 m, pushed in a river of 8 km/hr. Estimate the power required to overcome skin friction where:

$$C_f = \frac{0.0576}{{Re_x}^{0.2}} \qquad \tau_w = \frac{1}{2} C_f \rho V^2$$

Solution:

$$C_f = \frac{0.0576}{{Re_x}^{0.2}}$$

$$\therefore C_f = \frac{0.0576}{\left(\frac{\rho x U}{\mu}\right)^{0.2}} \qquad V = \frac{8 \times 100}{3600} = 2.22\ m/s \qquad x = L = 25\ m$$

$$C_f = \frac{0.0576}{\left(\frac{1000 \times 25 \times 2.22}{0.001}\right)^{6.2}} = 0.00163$$

$$\tau = \frac{1}{2} C_f . \rho V^2 = \frac{1}{2} \times 0.00163 \times 1000 \times (2.22)^2$$

$$= 4.02 \ N/m^2$$

$$F_1 = \tau A = 4.02 \times 25 \times 1.5 \times 2 = 301.35 \ N$$

$$F_2 = \tau A = 4.02 \times 25 \times 10 = 1004.5 \ N$$

$$F_T = F_1 + F_2 = 1305.86 \ N$$

$$Power = F.V = 2902 \ watt \ .$$

Q.19) .A laboratory wind tunnel has a lest section that is 3.5 mm square. Boundary-Layer velocity profiles are measured at two cross sections and displacement thickness are evaluated from the measured profile. At section (1):

$U_1 = 26 \ m/s$ $\delta^*{}_1 = 1.5 \ mm$

section (1):

$\delta^*{}_2 = 2.1 \ mm$

Calculate the change in static pressure between (1) & (2)

Solution:

$$\frac{P_1}{\rho g} + \frac{{U_1}^2}{2g} + z_1 = \frac{P_2}{\rho g} + \frac{{U_2}^2}{2g} + z_2$$

$$\therefore \ \frac{P_1 - P_2}{\rho g} = \frac{{U_2}^2}{2g} - \frac{{U_1}^2}{2g}$$

$$\therefore \ (P_1 - P_2) = \frac{\rho}{2} \left[{U_2}^2 - {U_1}^2\right]$$

$$P_1 - P_2 = \frac{\rho}{2} U_1^{\ 2} \left[\frac{U_2^{\ 2}}{U_1^{\ 2}} - 1\right]$$

$$\frac{P_1 - P_2}{\frac{1}{2}\rho U_1^{\ 2}} = \frac{U_2^{\ 2}}{U_1^{\ 2}} - 1 \qquad = \left(\frac{U_2}{U_1}\right)^2 - 1$$

$$U_1 A_1 = U_2 A_2 \qquad \Rightarrow \frac{U_2}{U_1} = \frac{A_1}{A_2}$$

$$\therefore \frac{P_1 - P_2}{\frac{1}{2}\rho U_1^{\ 2}} = \left(\frac{A_1}{A_2}\right)^2 - 1$$

$$A_1 = (0.305 - 2 \times 0.0015)^2$$

$$A_2 = (0.305 - 2 \times 0.0021)^2$$

$$\left(\frac{A_1}{A_2}\right)^2 - 1 = 0.0160531$$

$$\therefore P_1 - P_2 = \frac{1}{2}\rho U_1^{\ 2} \times 0.0160531$$

$$= \frac{1}{2} \times 1.2 \times 26^2 \times 0.0160531$$

$$= 6.51 \ N/m^2$$

Q.20). Using the velocity distribution

$$\frac{u}{U_\infty} = \sin\left(\frac{\pi}{2}\frac{y}{\delta}\right)$$

To determine the equation for growth of boundary layer thickness, shear stress, drag force and drag coefficient for laminar flow over flat plate.

Solution:

since $u = U_\infty \sin\left(\frac{\pi}{2}\eta\right)$ $\eta = \frac{y}{\delta}$

and

$$\therefore \quad \frac{du}{dy} = U_\infty \frac{\pi}{2\delta} . \cos\frac{\pi}{2}\frac{y}{\delta}$$

at $y = 0, \frac{du}{dy} = \frac{\pi}{2}\frac{U_\infty}{\delta} = 1.5708\frac{U_\infty}{\delta}$ **(1)**

$$\text{and} \quad \alpha = \int_0^1 \left(\sin\frac{\pi}{2}\eta - \sin^2\frac{\pi}{2}\eta\right) d\eta$$

$$= \frac{2}{\pi} - \frac{1}{2}\int_0^1 \sin^2\frac{\pi}{2}\eta \; d\eta$$

$$= \frac{2}{\pi} - \frac{1}{2}\int_0^1 \left(1 - \cos\frac{2\pi}{2}\eta\right) d\eta$$

$$= \frac{2}{\pi} - \left[\frac{\eta}{2}\right]_0^1 + \frac{1}{2\pi}\sin\pi\,\eta\Big]_0^1$$

$= \frac{2}{\pi} - \frac{1}{2} + 0 = 0.1366$ **(2)**

from eq. (1) $\tau = \mu \dfrac{du}{dy}$ at $y = 0$

$\therefore \quad \tau = 1.5708\frac{U_\infty}{\delta}$ **(3)**

and $\tau = \alpha\,\rho\,U_\infty^2 \dfrac{\partial\delta}{\partial x}$

$$= 0.1366\,\rho\,U_\infty^2 \frac{\partial \delta}{\partial x} \quad \textbf{(4)}$$

From eq. 3,4

$$0.1366\,\rho\,U_\infty^2 \frac{\partial \delta}{\partial x} = 1.5708 \frac{\mu U_\infty}{\delta}$$

or $$\delta\, d\delta = 11.5\,\frac{\mu}{\rho U_\infty}\,dx$$

$$\frac{\delta^2}{2} = 11.5\,\frac{\mu}{\rho U_\infty}\,x + C \qquad \begin{cases} B.C.\ \text{at}\ x = 0 \\ \qquad\quad \delta = 0 \\ \therefore \qquad C = 0 \end{cases}$$

or

$$\delta^2 = 23\,\frac{\mu\,x}{\rho\,U_\infty}$$

$$\frac{\delta^2}{x^2} = 23\,\frac{\mu}{\rho\,U_\infty\,x}$$

$$\therefore \qquad \frac{\delta}{x} = \frac{4.8}{\sqrt{Re_x}} \quad \textbf{(5)}$$

Then

$$\tau = \frac{\mu\,U_\infty}{\delta} \times 1.5708$$

$$= 1.5708 \times \mu\,U_\infty \Big/ \sqrt{23\,\frac{\mu\,x}{\rho\,U_\infty}}$$

$$\tau = 0.327\sqrt{\frac{\rho\,\mu\,U_\infty^3}{x}}$$

$$F_D = \tau A = \int_0^L \tau\,dx \times 1 \quad \textbf{(6)}$$

$$= 0.327\sqrt{\rho\,\mu\,U_\infty^3} \times \int_0^L x^{-1/2}\,dx$$

$$F_D = 0.654\sqrt{\rho\,\mu\,U_\infty^3\,L} \qquad \textbf{(7)}$$

$$C_D = \frac{F_D}{\frac{1}{2}\rho\,U_\infty^2\,L} = \frac{0.654\sqrt{\rho\,\mu\,U_\infty^3\,L}}{\frac{1}{2}\rho\,U_\infty^2\,L}$$

$$C_D = \frac{1.308}{\sqrt{Re_L}} \qquad \textbf{(8)}$$

Q.21). For the velocity distribution

$$\frac{u}{U_\infty} = \sin\left(\frac{\pi}{2}\frac{y}{\delta}\right)$$

Determine the displacement thickness, momentum thickness, and the skin friction coefficient C_f ?

Solution:

$$\text{since} \quad \delta = \frac{4.8}{\sqrt{Re_x}}$$

$$\text{Then} \quad \delta^* = \int_0^1 \left(1 - \sin\frac{\pi}{2}\eta\right) d\eta \qquad \eta = \frac{y}{\delta}$$

$$= \delta\left[\eta + \frac{2}{\pi}\cos\frac{\pi}{2}\eta\right]_0^1$$

$$\delta^* = \delta\left[1 + 0 - \frac{2}{\pi}\right] = 0.363\,\delta$$

or $$\frac{\delta^*}{x} = \frac{1.743}{\sqrt{Re_x}} \quad \textbf{(1)}$$

and $$\theta = \alpha\delta$$

$\alpha = 0.1366$ from previous question

$$\therefore \quad \frac{\theta}{x} = \frac{0.655}{\sqrt{Re_x}} \quad \textbf{(2)}$$

$$C_f = \frac{\theta}{x} = \frac{0.655}{\sqrt{Re_x}} \quad \textbf{(3)}$$

Q.22). Drive the turbulent boundary thickness equation based on the exponential law

$$\frac{u}{U_\infty} = \left(\frac{y}{\delta}\right)^{\frac{1}{9}} \quad \text{for} \quad f = 0.185/Re^{0.2}$$

and $$\tau_\circ = \frac{f}{8}\,\rho\,V^2$$

Solution:

$$\frac{u}{U_\infty} = \left(\frac{y}{\delta}\right)^{\frac{1}{9}} = \eta^{\frac{1}{9}} \qquad \eta = \frac{y}{\delta}$$

$$\alpha = \int_0^1 \eta^{\frac{1}{9}}\left(1 - \eta^{\frac{1}{9}}\right) d\eta$$

$$= \frac{9}{110}$$

$$\tau_\circ = \frac{9}{110}\,\rho\,U_\infty^2\,\frac{\partial\delta}{\partial x} \quad \textbf{(1)}$$

and

$$\tau_\circ = C_f \frac{1}{2} \rho V^2 = \frac{f}{8} \rho V^2$$

$$\tau_\circ = \frac{0.185}{(Re_D)^{0.2}} \cdot \frac{1}{8} \rho V^2$$

$$\tau_\circ = \frac{0.185}{\sqrt{\frac{\rho D V}{\mu}}} \cdot \frac{1}{8} \rho V^2$$

Since $\quad V = \dfrac{U_\infty}{1.235} \qquad$ when $\quad D = 2\delta \quad$ (streeter)

Then

$$\tau_\circ = \frac{0.185}{\left(\frac{\rho 2\delta U_\infty}{\mu \times 1.235}\right)^{0.2}} \cdot \frac{1}{8} \rho U_\infty^2 \times \frac{1}{(1.235)^2}$$

$$\tau_\circ = \rho U_\infty^2 \times \frac{0.01377}{(Re_\delta)^{0.2}} \qquad \textbf{(2)}$$

From eq. 1,2 we get

$$\frac{9}{110} \rho U_\infty^2 \frac{\partial \delta}{\partial x} = 0.01377 \times \rho U_\infty^2 \times \left(\frac{\mu}{\rho U_\infty}\right)^{0.2} \times \frac{1}{\delta^{0.2}}$$

or $\quad \delta^{1/5} \partial\delta = 0.1683 \left(\dfrac{\mu}{\rho U_\infty}\right)^{1/5} \partial x$

or $\quad \dfrac{\delta}{x} = 0.2673 (Re_x)^{1/6}$

Q.23). Air flows a long smooth plate with a velocity of 40 m/s. How long the plate have to obtain a boundary – layer thickness of 8 mm ? when the kinematic viscosity of air $1.6 \times 10^{-5}\ m^2/s$.

Solution:

Assume T.B.L. and

$$Re = \frac{X\,V}{\upsilon} = \frac{40\,X}{1.6 \times 10^{-5}}$$

$$Re = 25 \times 10^{5} X$$

Since

$$\delta = \frac{0.37\,X}{(Re)^{0.2}} = 0.008$$

$$0.37\,X = (25 \times 10^{5}) \times X^{0.2} \times 0.008$$

$$X = 0.412\,X^{0.2}$$

$$X^{0.8} = 0.412$$

$$X = 0.33\;m$$

at $X = L$

$$\therefore\;\; L = 33\;cm$$

Q.24). The walls of a wind tunnel are sometimes made divergent to offset the effect of the boundary layer in reducing the portion of the cross section in which the flow is of constant speed. At what angle must plane walls be set so that the displacement thickness does not encroach upon the tunnel's constant – speed cross section at distance greater than 24 cm from the leading edge of the wall? $\upsilon = 1.6 \times 10^{-5}\; m^2/s$,$V = 40\; m/s$

Solution:

$$V = 40\; m/s, \quad \upsilon = 1.6 \times 10^{-5}\; m^2/s$$

Assume T.B.L. and

$$Re = \frac{X\,V}{\upsilon} = \frac{X40}{1.6 \times 10^{-5}} = 25 \times 10^{5} X$$

and

$$\delta = \frac{0.37\,X}{(Re)^{0.2}}$$

$$\delta = \frac{0.37\,X}{(25 \times 10^5 X)^{0.2}} = 0.0194\ X^{0.8}$$

$$\frac{\partial \delta}{\partial x} = 0.194\ \times 0.8\,X^{-0.2}$$

at $X = 24\,cm = 0.24\,m$

$$\frac{\partial \delta}{\partial x} = \frac{0.01555}{(0.24)^{0.2}} = 0.0207$$

$$\theta = \tan^{-1}\frac{\partial \delta}{\partial x} = 1.185° \quad \text{angle of slope}$$

Q.25). the general equation of shear stress over flat plate as follows:

$$\tau_w = \frac{\partial}{\partial x}\int_0^{\delta} \rho\, u(U-u)dy + \frac{dU}{dx}\int_0^{\delta} \rho\,(U-u)dy$$

Show that

$$\frac{\tau_w}{\rho U^2} = \frac{\partial \theta}{\partial x} + \frac{1}{U}\frac{dU}{dx}\theta[2+H]\,.$$

Solution:

- for the laminar flow over flat plate the shear stress at the wall state:

$$\tau_w = \frac{\partial}{\partial x}\int_0^{\delta} \rho\, u(U-u)dy + \frac{dU}{dx}\int_0^{\delta} \rho\,(U-u)dy$$

Where:

$U = free\ stream\ velocity.$

$u = flow\ in\ side\ the\ B.L.$

$\theta = momentum\ thickness.$

$\delta = B.L. thickness.$

Now:

$$\tau_w = \frac{\partial}{\partial x}\int_0^\delta \rho\, U^2 \frac{u}{U}\left(1 - \frac{u}{U}\right) dy + \frac{1}{U}\frac{dU}{dx}\int_0^\delta \rho\, U^2 \left(1 - \frac{u}{U}\right) dy$$

$\therefore$ Since. $\theta = momentum\ thickness = \int_0^\delta \frac{u}{U}\left(1 - \frac{u}{U}\right) dy$

& $\delta^* = displacemat\ thickness = \int_0^\delta \left(1 - \frac{u}{U}\right) dy$

$$\therefore \quad \tau_w = \frac{\partial}{\partial x}(\rho\, U^2\, \theta) + \frac{1}{U}\frac{dU}{dx}\delta^*$$

$*$ Since the shape factor $H = \frac{\delta^*}{\theta}$

$\therefore \quad \delta^* = H\theta$

$$\therefore \quad \tau_w = \frac{\partial}{\partial x}(\rho\, U^2\, \theta) + \frac{1}{U}\frac{dU}{dx} H\theta$$

$$\therefore \quad \tau_w = \rho\, U^2 \frac{\partial \theta}{\partial x} + 2\, \rho\, U\, \theta \frac{dU}{dx} + \frac{1}{U}\frac{dU}{dx} H\theta\,.$$

$$\therefore \quad \tau_w = \rho\, U^2 \left[\frac{\partial\theta}{\partial x} + \frac{1}{U}\frac{dU}{dx}2\theta + \frac{1}{U}\frac{dU}{dx}H\theta\right].$$

$$\therefore \quad \tau_w = \rho\, U^2 \left[\frac{\partial\theta}{\partial x} + \frac{1}{U}\frac{dU}{dx}\theta + (2+H)\right].$$

$$\therefore \quad \frac{\tau_w}{\rho\, U^2} = \frac{\partial\theta}{\partial x} + \frac{\theta}{U}\frac{dU}{dx}(2+H).$$

$*$ For the flow over a flat plate with. $\dfrac{dU}{dx} = 0$.

$\& \ \dfrac{dP}{dx} = 0$.

$$\therefore \quad \frac{\tau_w}{\rho\, U^2} = \frac{\partial\theta}{\partial x}$$

Q.26.).For Laminar B.L. the velocity profile

$\dfrac{U}{U_\infty} = 2\eta - \eta^2$ evaluate $\delta, \delta^*, \theta, \delta^{**}, C_f, C_D$ in terms of Re. No.

$$u = U_\infty(2\eta - \eta^2)$$

$$\frac{du}{dy} = U_\infty\left(\frac{2}{\delta} - \frac{2y}{\delta^2}\right) \quad \text{at} \quad y = 0 \qquad \frac{du}{dy} = \frac{2U_\infty}{\delta}$$

$$\tau_0 = \rho U_\infty^2 \frac{\partial\delta}{\partial x}\int_0^1 f(\eta)\,(1 - f(\eta))d\eta = \rho U_\infty^2 \frac{\partial\delta}{\partial x}\alpha$$

$$\therefore \quad \alpha = \int_0^1 (2\eta - \eta^2)\,(1 - 2\eta + \eta^2)d\eta$$

$$= \int_0^1 (2\eta - 5\eta^2 + 4\eta^3 - \eta^4)\, d\eta = \left[\eta^2 - \frac{5}{3}\eta^3 + \eta^4 - \frac{\eta^5}{5}\right]_0^1$$

$$\alpha = 0.133 \qquad\qquad = \left[1 - \frac{5}{3} + 1 - \frac{1}{5}\right] = 0.1333$$

$$\therefore \quad \mu \frac{2U_\infty}{\delta} = 0.1333\, \rho U_\infty^2 \frac{\partial \delta}{\partial x}\Bigg] \div \partial\delta$$

$$\delta\, \partial\delta = \frac{15\,\mu}{\rho U_\infty} dx$$

$$\frac{\delta^2}{2} = 15 \frac{\mu x}{\rho U_\infty} + \text{Const} \quad \text{at} \quad x = 0 \quad \delta = 0 \quad \Rightarrow \text{Const} = 0$$

$$\delta = \frac{5.477\, x}{\sqrt{Re_x}}$$

$$\delta^* = \delta \int_0^1 (1 - f(\eta))\, d\eta = \delta \int_0^1 (1 - 2\eta + \eta^2) d\eta = \delta \left[\eta - \eta^2 + \frac{\eta^3}{3}\right]_0^1$$

$$\therefore \quad \delta^* = \frac{\delta}{3} = \frac{1.8256\, x}{\sqrt{Re_x}}$$

$$\theta = \delta \int_0^1 f(\eta)(1 - f(\eta))\, d\eta = \alpha\delta = \frac{0.73\, x}{\sqrt{Re_x}}$$

$$\delta^{**} = \delta \int_0^1 (f(\eta) - f^3(\eta))\, d\eta$$

$$= \delta \int_0^1 \left((2\eta - \eta^2) - (2\eta - \eta^2)^3\right) d\eta$$

$$= \delta \left[\eta^2 - \frac{\eta^3}{3} - 2\eta^4 + \frac{12}{5}\eta^5 - \eta^6 + \frac{\eta^7}{7} \right]_0^1 = 0.2095\, \delta$$

$$= \frac{1.147\, x}{\sqrt{Re_x}}$$

$$C_f = \frac{\theta}{x} = \frac{0.73\, x}{\sqrt{Re_x}}$$

$$C_F = 2C_f = \frac{1.46\, x}{\sqrt{Re_x}}$$

$$\tau_0 = \mu \left. \frac{du}{dy} \right|_{y=0} = \frac{2U_\infty\, \mu}{\delta} = \frac{2}{5.477} \sqrt{Re_x} \,. \frac{\mu U_\infty}{x}$$

$$= 0.365\, \frac{\rho U_\infty^2}{\sqrt{Re_x}}$$

or

$$A = \alpha = \frac{2}{15} \qquad \text{from integration} \quad \int_0^1 f(\eta)(1 - f(\eta))\, d\eta$$

Since

$$\left. \frac{df(\eta)}{d\eta} \right|_{\eta=0} = \mathrm{B} = 2 - 2\eta = 2$$

$$\delta = \sqrt{\frac{2B}{A}}\, \frac{x}{\sqrt{Re_x}}$$

$$\theta = A\, \delta$$

$$\tau_0 = \frac{\rho U_\infty^2}{\sqrt{Re_x}}\sqrt{\frac{AB}{2}}$$

one side $F = \int_0^\ell \tau_0 \, dx = \sqrt{2AB\mu\rho U_\infty^3 \ell}$

$$C_F = \frac{2\sqrt{2AB}}{\sqrt{Re_L}}$$

Q.27) . For laminar B.L. the velocity profile

$$f(\eta) = \frac{u}{U_\infty}$$

$= \frac{3}{2}\eta - \frac{1}{2}\eta^3$ evaluate $\delta, \delta^*, \theta, C_f, \tau_0$ as a function of Re. No.

$$\left.\frac{df(\eta)}{d\eta}\right|_{\eta=0} = \left(\frac{3}{2}\eta - \frac{1}{2}\eta^3\right) = \frac{3}{2} = B$$

$$\tau_0 = \rho U_\infty^2 \frac{\partial \delta}{\partial x}\int_0^1 f(\eta)\left(1 - f(\eta)\right)d\eta = \rho U_\infty^2 \frac{\partial \delta}{\partial x}\alpha$$

$$\alpha = \int_0^1 \left(\frac{3}{2}\eta - \frac{1}{2}\eta^3\right)\left(1 - \frac{3}{2}\eta - \frac{1}{2}\eta^3\right)d\eta$$

$$= \int_0^1 \left(\frac{3}{2}\eta - \frac{9}{4}\eta^2 + \frac{3}{4}\eta^4 - \frac{1}{2}\eta^3 + \frac{3}{4}\eta^4 - \frac{1}{4}\eta^6\right)d\eta$$

$$= \int_0^1 \left(\frac{3}{2}\eta - \frac{9}{4}\eta^2 + \frac{6}{4}\eta^4 - \frac{1}{2}\eta^3 - \frac{1}{4}\eta^6\right)d\eta$$

$$= \left[\frac{3}{4}\eta^2 - \frac{3}{4}\eta^3 - \frac{1}{8}\eta^4 + \frac{6}{20}\eta^5 - \frac{1}{28}\eta^7\right]_0^1$$

$$= \left[\frac{3}{4} - \frac{3}{4} - \frac{1}{8} + \frac{6}{20} - \frac{1}{28}\right] = 0.13928 = A$$

$$\delta = \sqrt{\frac{2B}{A}}\frac{x}{\sqrt{Re_x}} = \frac{4.64\,x}{\sqrt{Re_x}}$$

$$\delta^* = \delta\int_0^1 \left(1 - \frac{3}{2}\eta - \frac{1}{2}\eta^3\right) d\eta = \frac{4.64\,x}{\sqrt{Re_x}}\left[\eta - \frac{3}{4}\eta^2 + \frac{1}{8}\eta^3\right]_0^1 = 0.375\,\delta$$

$$\delta^* = \frac{1.74\,x}{\sqrt{Re_x}}$$

$$\theta = A.\delta = 0.13928 \times \frac{4.64\,x}{\sqrt{Re_x}} = \frac{0.646\,x}{\sqrt{Re_x}}$$

$$C_f = \frac{\theta}{x} = \frac{0.646}{\sqrt{Re_x}}$$

$$\tau_0 = \sqrt{\frac{AB}{2}}\frac{\rho U_\infty^2}{\sqrt{Re_x}} = \frac{0.323\,\rho U_\infty^2}{\sqrt{Re_x}} = \frac{C_f\,\rho U_\infty^2}{2}$$

Q.28). For laminar B.L. the velocity profile

$$\frac{U}{U_\infty} = 2\eta - 2\eta^3 + \eta^4 \quad \text{evaluate } \delta, \delta^*, \theta, C_f, \tau_0 \quad \text{as a finction Re. No.}$$

$$\frac{u}{U_\infty} = (2\eta - 2\eta^3 + \eta^4) = f(\eta)$$

$$\left.\frac{df(\eta)}{d\eta}\right|_{\eta=0} = 2 - 6\eta^2 + 4\eta^3|_{\eta=0} = 2 = B$$

$$\alpha = \int_0^1 (2\eta - 2\eta^3 + \eta^4) - (2\eta - 2\eta^3 + \eta^4)^2 d\eta$$

$$= \int_0^1 (2\eta - 2\eta^3 + \eta^4 - 4\eta^2 + 8\eta^4 - 4\eta^5 - 4\eta^6 + 4\eta^7 - \eta^8)\, d\eta$$

$$= \left[\eta^2 - \frac{1}{2}\eta^4 + \frac{\eta^5}{5} - \frac{4}{3}\eta^3 + \frac{8}{5}\eta^5 - \frac{4}{6}\eta^6 - \frac{4}{7}\eta^7 + \frac{1}{2}\eta^8 - \frac{\eta^9}{9}\right]_0^1$$

$$= 0.11746 = A$$

$$\delta = \sqrt{\frac{2B}{A}} \frac{x}{\sqrt{Re_x}} = \frac{5.836\, x}{\sqrt{Re_x}}$$

$$\delta^* = \delta \int_0^1 (1 - 2\eta + 2\eta^3 - \eta^4)\, d\eta = \delta \left[\eta - \eta^2 + \frac{1}{2}\eta^4 - \frac{\eta^5}{5}\right]_0^1$$

$$= 0.3\, \delta = \frac{1.7507\, x}{\sqrt{Re_x}}$$

$$\theta = \alpha\delta = \frac{0.685\, x}{\sqrt{Re_x}}$$

$$C_f = \frac{\theta}{x} = \frac{0.685}{\sqrt{Re_x}}$$

$$\tau_0 = \sqrt{\frac{AB}{2}} \frac{\rho U_\infty^2}{\sqrt{Re_x}} = \frac{C_f}{2}\, \rho U_\infty^2 = 0.3427 \frac{\rho U_\infty^2}{\sqrt{Re_x}}$$

$$\delta^{**} = \frac{1.02\,x}{\sqrt{Re_x}}$$

Q.29). For laminar B.L. the velocity profile

$$\frac{u}{U_\infty} = \sin\left(\frac{\pi}{2}\eta\right)$$

evaluate $\delta, \delta^*, \theta, C_f, \tau_0$ as a finction Re. No. and the drag force

where $\int \sin^2 x = \frac{1}{2}(1 - \cos 2x)dx$

$$f(\eta) = \frac{u}{U_\infty} = \sin\left(\frac{\pi}{2}\eta\right)$$

$$\left.\frac{df(\eta)}{d\eta}\right|_{\eta=0} = \frac{\pi}{2}\cos\left(\frac{\pi}{2}\eta\right)\Big|_{\eta=0} = \frac{\pi}{2} = B$$

$$\alpha = \int_0^1 \left(\sin\frac{\pi}{2}\eta - \sin^2\frac{\pi}{2}\eta\right)\,d\eta = \frac{2}{\pi}\left[-\cos\frac{\pi}{2}\eta\right]_0^1 - \int_0^1 \sin^2\frac{\pi}{2}\eta\;d\eta$$

$$= \frac{2}{\pi} - \frac{1}{2}\int_0^1 \left(1 - \cos\frac{2\pi}{2}\eta\right)\,d\eta = \frac{2}{\pi} - \left[\frac{\eta}{2}\right]_0^1 + \left.\frac{1}{2\pi}\sin\pi\eta\right]_0^1$$

$$= \frac{2}{\pi} - \frac{1}{2} = 0.1366 = A$$

$$\delta = \sqrt{\frac{2B}{A}}\frac{x}{\sqrt{Re_x}} = \frac{4.796\,x}{\sqrt{Re_x}}$$

$$\delta^* = \delta \int_0^1 \left(1 - \sin\frac{\pi}{2}\eta\right) d\eta = \delta \left[\eta + \frac{2}{\pi}\cos\frac{\pi}{2}\eta\right]_0^1 = \delta\left(1 - \frac{2}{\pi}\right)$$

$$= \frac{1.743\, x}{\sqrt{Re_x}}$$

$$\theta = \alpha\delta = \frac{0.655\, x}{\sqrt{Re_x}}$$

$$C_f = \frac{\theta}{x} = \frac{0.655}{\sqrt{Re_x}}$$

$$\tau_0 = \frac{C_f}{2}\, \rho U_\infty^2 = 0.3275 \frac{\rho U_\infty^2}{\sqrt{Re_x}}$$

Cheek $\quad D = \int_0^L \tau_0\, dx = \int_0^L 0.3275 \frac{\rho U_\infty^2}{\sqrt{Re_x}} dx$

$$= \int_0^L 0.3275\, \mu U_\infty \left(\frac{\rho U_\infty}{\mu\, x}\right)^{\frac{1}{2}} dx$$

$$= 0.654\, \rho U_\infty^2\, L / \left(\frac{\rho U_\infty L}{\mu}\right)^{\frac{1}{2}}$$

Q.30). Given $f = \frac{0.185}{(Re)^{1/5}}$, $\tau_0 = \frac{\rho V^2 f}{8}$, $U = 1.23\, v$

$$C_F = \frac{0.37}{(Re)^{1/5}}$$

$$C_F \alpha Re^{-2/(n+1)} \quad \Rightarrow n = 9$$

$\frac{U}{U_\infty} = \left(\frac{y}{\delta}\right)^{1/9} = (\eta)^{1/9}$ evaluate $\delta, \delta^*, \theta, \delta^{**}$

$$\alpha = \int_0^1 f(\eta)\,(1 - f(\eta))d\eta = \int_0^1 \left(\eta^{\frac{1}{9}} - \eta^{\frac{2}{9}}\right) d\eta$$

$$= \frac{9}{10}\eta^{10/9} - \frac{9}{11}\eta^{11/9}\Big]_0^1 \frac{9}{10} - \frac{9}{11} = \frac{9}{110}$$

$$\tau_0 = \frac{\rho V^2}{8} \times \frac{0.185}{(Re)^{\frac{1}{5}}} = \frac{\rho U_\infty^2}{8 \times (1.23)^2} \times \frac{0.185}{(Re)^{\frac{1}{5}}} = 0.01528\,\rho U_\infty^2 \,.\left(\frac{\upsilon}{U_\infty \delta}\right)^{1/5}$$

$$\frac{9}{110}\rho U_\infty^2 \frac{\partial \delta}{\partial x} = 0.01528\,\rho U_\infty^2 \,.\left(\frac{\upsilon}{U_\infty}\right)^{\frac{1}{5}} .\left(\frac{1}{\delta}\right)^{\frac{1}{5}}$$

$$\delta^{\frac{1}{5}}\,\partial\delta = 0.1868\left(\frac{\upsilon}{U_\infty}\right)^{\frac{1}{5}} dx$$

$$\frac{5}{6}\delta^{\frac{6}{5}} = 0.1868\left(\frac{\upsilon}{U_\infty}\right)^{\frac{1}{5}} x + \text{Const} \qquad x = 0 \quad \delta = 0$$

$$\delta^{\frac{6}{5}} = 0.224\left(\frac{\upsilon}{U_\infty}\right)^{\frac{1}{5}} x$$

$$\delta = \frac{0.287\,x}{(Re_x)^{\frac{1}{6}}}$$

$$\delta^* = \delta\int_0^1 \left(1 - \eta^{\frac{1}{9}}\right) d\eta = \delta\left[\eta + \frac{9}{10}\eta^{\frac{10}{9}}\right]_0^1 = 0.1\,\delta$$

$$= \frac{0.287\,x}{(Re_x)^{\frac{1}{6}}}$$

$$\theta = \alpha\delta = \frac{9}{110} \times \frac{0.287\, x}{(Re_x)^{\frac{1}{6}}} = \frac{0.023\, x}{(Re_x)^{\frac{1}{6}}}$$

$$\delta^{**} = \delta \int_0^1 \left(\eta^{\frac{1}{9}} - \eta^{\frac{3}{9}}\right) d\eta$$

$$= \delta \left[\frac{9}{10}\eta^{\frac{10}{9}} - \frac{9}{12}\eta^{\frac{12}{9}}\right]_0^1 = 0.15\, \delta = \frac{0.043\, x}{(Re_x)^{\frac{1}{6}}}$$

Q.31). Prove that Blasius equation for the following assumptions. (isothermal – steady – incompressible, laminar B.L. on a flat plate)

$$u\frac{\partial u}{\partial x} + \mathrm{v}\frac{\partial u}{\partial y} = \mathrm{v}\frac{\partial^2 u}{\partial x^2} \qquad \text{momentum eq.} \qquad \frac{dP}{dx} = 0$$

$$\frac{\partial u}{\partial x} + \frac{\partial v}{\partial y} = 0$$

$$at \;\; y = 0 \qquad u = \mathrm{v} = 0$$

$$at \;\; y = \infty \qquad U = U_\infty$$

$$\frac{u}{U_\infty} = \emptyset\left(\frac{y}{\delta}\right), \qquad \delta \propto \frac{x}{\sqrt{Re_x}} \quad \Rightarrow \quad \delta \propto \sqrt{\frac{\upsilon x}{U_\infty}}$$

$$\frac{u}{U_\infty} = \emptyset\left(\frac{y}{\sqrt{\frac{\upsilon x}{U_\infty}}}\right) = \emptyset\left(y\sqrt{\frac{\upsilon x}{U_\infty}}\right) = \emptyset(\eta)$$

$$\varphi = \sqrt{xU_\infty \upsilon}\, f(\eta)$$

$$u = \frac{\partial \varphi}{\partial y} = \frac{\partial \varphi}{\partial \eta}.\frac{\partial \eta}{\partial y} = \sqrt{xU_\infty \upsilon}\, f'(\eta).\sqrt{\frac{U_\infty}{\upsilon x}} = U_\infty f'(\eta)$$

$$\therefore \frac{u}{U_\infty} = f'(\eta) = \emptyset(\eta) \quad \text{where} \quad f'(\eta) = \frac{\partial f(\eta)}{\partial \eta}$$

$$\mathrm{v} = \frac{-\partial \varphi}{\partial x} = -\left\{\sqrt{xU_\infty \upsilon}\, f'(\eta)\frac{\partial \eta}{\partial x} + f(\eta)\frac{\partial}{\partial x}\sqrt{xU\upsilon}\right\}$$

$$= -\left\{\sqrt{xU_\infty \upsilon}\, f'(\eta).y \times \left(-\frac{1}{2}\sqrt{\frac{U_\infty}{\upsilon x^3}}\right) + \frac{1}{2}\sqrt{\frac{U_\infty \upsilon}{x}} f(\eta)\right\}$$

$$= \frac{1}{2}\sqrt{\frac{U_\infty \upsilon}{x}}\{ f'(\eta).\eta - f(\eta)\}$$

$$\frac{\partial u}{\partial x} = \frac{\partial}{\partial x} U_\infty f''(\eta)\frac{\partial \eta}{\partial x}$$

$$= U_\infty .y\left(-\frac{1}{2}\sqrt{\frac{U_\infty}{\upsilon x^3}}\right) f''(\eta) = -\eta\frac{U}{2x} f''(\eta)$$

$$\frac{\partial u}{\partial y} = \frac{\partial}{\partial \eta} U_\infty f(\eta)\frac{\partial \eta}{\partial y} = U_\infty f''(\eta).\sqrt{\frac{U_\infty}{\upsilon x}}$$

$$\frac{\partial^2 u}{\partial y^2} = U_\infty \sqrt{\frac{U_\infty}{\upsilon x}} f'''(\eta).\frac{\partial \eta}{\partial y} = \frac{U_\infty^2}{\upsilon x} f'''(\eta)$$

Substuting for $u, \mathrm{v}, \frac{\partial u}{\partial x}, \frac{\partial u}{\partial y} \& \frac{\partial^2 u}{\partial y^2}$ in momentum eqn.

$$-U_\infty f'(\eta)\frac{\eta U_\infty}{2x} f''(\eta) + \frac{1}{2}\sqrt{\frac{\upsilon x}{x}} \{\eta f'(\eta) - f(\eta)\}\, U_\infty \sqrt{\frac{U_\infty}{\upsilon x}} f''(\eta)$$

$$= \frac{U_\infty^2}{x} f'''(\eta) \times \frac{x}{U_\infty^2}$$

$$-\frac{f'(\eta).\eta.f''(\eta)}{2} + \frac{\eta f'(\eta)f''(\eta)}{2} - \frac{f(\eta)f''(\eta)}{2} = f'''(\eta)$$

$\therefore\ f(\eta)f''(\eta) + 2f'''(\eta) = 0$ (Blasius eqn)

Q.32)A uniform free stream of air at 10 m/sec flows over a flat plate. Calculate the drag coefficient and the drag for the plate, 0.5 m long and 2 m wide.

Take $\rho = 1.2\ kg/m^3, \mu = 18 \times 10^{-6}\ N.s/m^2$

$$Re_L = \frac{\rho U_\infty L}{\mu} = \frac{1.2 \times 0.5 \times 10}{18 \times 10^{-6}} = 3.33 \times 10^5 < 5 \times 10^5 \quad \text{Laminar flow}$$

$$C_D = \frac{1.328}{\sqrt{Re_L}} = 2.3 \times 10^{-3}$$

$$D = 0.5 \times \rho \times U_\infty^2 \times C_D \times B \times L$$

$$= 0.5 \times\times 1.2 \times 100 \times 0.0023 \times 0.5 \times 2$$

$$= 0.138\ N$$

Q.33). Calculate the drag and power required to a smooth flat plate 2 m wide, 20 m long through still water $\mu = 0.001\ N.s/m^2\ at\ 10\ m/s$.

$$Re_L = \frac{\rho U_\infty L}{\mu} = \frac{1000 \times 10 \times 20}{0.001} = 2 \times 10^8 > 5 \times 10^5 \quad \Rightarrow \text{Turbulent}$$

for $Re_L > 10^7$ use the following relationship. For C_D calculation.

$$C_D = \frac{0.455}{(\log_{10} Re_L)^{2.58}} = 0.0019348$$

$$D = \frac{1}{2}\rho U_\infty^2 C_D\, A$$

$$= \frac{1}{2} \times 1000 \times 10^2 \times 0.0019348 \times 2 \times 20$$

$$= 3870\ N$$

For both sides $D = 7740\ N$

power required $= D \times V = 7740 \times 10 = 77.4\ kW$

Q.34). Calculate the friction drag on a plate 0.15 m wide and 0.45 m long placed longitudinally in a stream of oil ($\rho = 925\ kg/m^3, \upsilon = 9 \times 10^{-5}\ m^2/s$) flowing with a free stream velocity of 6 meters per second. Also find the thickness of the boundary layer and shear stress at the trailing edge.

$$Re_L = \frac{\rho U_\infty L}{\mu} = 30000 < 5 \times 10^5$$

$\therefore$ Laminar flow

$$C_D = 1.328/\sqrt{Re_L} = 0.0077$$

$$D = \frac{1}{2}\rho U_\infty^2 C_D\, A = \frac{1}{2} \times 925 \times (6)^2 \times (0.45 \times 0.8) \times 0.0077$$
$= 8.65\ N$ for one surface

for both sides $D = 17.3\ N$.

$$\delta_{TE} = \frac{4.91\ L}{\sqrt{Re_L}} = \frac{4.91\ \times 0.5}{\sqrt{30000}} = 0.0127\ m\ \simeq 13\ mm$$

$$\tau_0 = \frac{1}{2}\rho U^2 C_F = \frac{1}{2}\rho U^2 \frac{C_D}{2} = 0.5 \times 925 \times 36 \times \frac{0.0077}{2}$$

$$\simeq 63\ N/m^2$$

Q.35). If $\frac{u}{U_\infty} = 2\eta - \eta^2$, find the thickness of the B.L., the shear stress at the T.E, and the drag force on one side of the plate 1 m long if it is immersed in water flowing with a velocity of 0.3 m/s.

$$\delta_{TE} = \frac{5.478 \times 1}{\sqrt{3 \times 10^5}} = 0.01\, m = 10\, mm$$

$$\tau_{0TE} = \frac{2\mu U_\infty}{\delta} = \frac{2 \times 0.001 \times 0.3}{0.01} = 0.06\ N/m^2$$

$$D = \frac{0.73\, \rho U^2 L}{\sqrt{Re_L}} = 0.1199\, N$$

Q36). A flat plate 1 m square is held normal to the flow of air at 6 m/s. It is found to experience a drag force of 200 N. Determine the drag force when the plate is held parallel to the flow. Find the rate of two drags. Assume ($\rho = 1.2\ kg/m^3, \upsilon = 1.5 \times 10^{-5}\ m^2/s$) further assume the boundary layer to be turbulent.

Solution:

$$Re_L = \frac{\rho U_\infty L}{\mu} = \frac{1.2 \times 6 \times 1}{1.5 \times 10^{-5}} = 4 \times 10^5$$

$$Re_L < 10^7$$

$$C_D = \frac{0.072}{(Re)^{1/5}} = 0.00546$$

$$D = \frac{1}{2}\rho U_\infty^2 C_D\, A \times 2$$

$$= 0.236\, N$$

$$D_{ratio} = \frac{0.24}{200} = 1.2 \times 10^{-3}$$

Q.37). A smooth flat plate 2.4 m long and 0.9 m wide moves length ways at 6 m/s through still atmospheric air of density 1.2 kg/m^3 and kinematic viscosity

$14.9\ mm^2/s$. Assuming the boundary layer to be entirely laminar. Calculate the boundary layer thickness at the trailing edge of the plate, the shear stress half way a long and the power required to move the plate. What power would be required if the boundary layer were made turbulent at the leading edge?

Solution:

$$\delta_{TE} = \frac{4.91\,x}{\sqrt{Re_x}} = \frac{4.91\ \times 2.4}{\left(\frac{6 \times 2.4}{14.9 \times 10^{-6}}\right)^{1/5}} = 0.01199\ m = 11.99\ mm$$

$$\tau_{0_{1/2}} = \frac{1}{2}\rho U_\infty^2 C_F = 0.5 \times 1.21 \times 36 \times \frac{0.664}{\sqrt{\frac{6 \times 1.2}{14.9 \times 10^{-6}}}} = 0.0208\ N/m^2$$

$$\text{Power required} = F_x \times U_\infty = 0.5\ \rho U_\infty^2\ AC_D\ \times 2 \times U_\infty$$

$$= 1.21 \times (6)^3 \times 2.4 \times 0.9 \frac{1.328}{\sqrt{\frac{6 \times 2.4}{14.9 \times 10^{-6}}}} = 0.763\ W$$

$$\text{Power required})_{T\ at\ LE} = 1.21 \times (6)^3 \times 0.9 \frac{0.072}{\sqrt{\frac{6 \times 2.4}{14.9 \times 10^{-6}}}} = 2.636\ W$$

Q.38). A stream lined vehicle is 10 m long, with sides 2.75 m high and the top2.75 m wide. Assuming the skin – friction drag on sides and top to be equal to that on one side of a flat 8.25 m wide and 10 m long, compute the power required to overcome the skin – friction drag when the vehicle is traveling at 110 km/hr through standard atmosphere air at sea level. Assume that the boundary layer is disturbed near the leading edge.

Solution:

$$U_\infty = 110 \times 1000/3600 = 30.56\ m/s$$

$\rho = 1.2\ kg/m^3, \upsilon = 18 \times 10^{-6}\ m^2/s$

$$Re_L = \frac{\rho U_\infty L}{\mu} = \frac{1.2 \times 10 \times 30.56}{18 \times 10^{-6}} = 2 \times 10^7$$

$$P = D \times V = 0.5 \times 1.2 \times (30.56)^2 \times 10 \times 8.25 \times \frac{0.455}{\log_{10}(2 \times 10^2)^{2.28}}$$

$= 3806.7\ W$

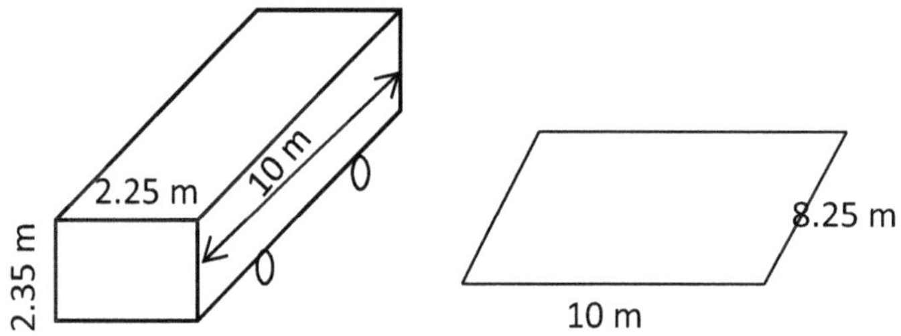

Q.39). Air at 20⁰ C, and 100 kpa abs flow a long rectangular smooth flat plate with 0.3 m width and 1 m length. Evaluate shear stress at the mid of the plate when $Re_x = UX/\upsilon = 10^6$. And the skin friction drag for the whole plate. If the flow is completely turbulent over the plate. The well-known approximate formula for the wall friction coefficient is

Solution:

$C_f = 0.059/Re_x^{\frac{1}{5}}, \upsilon_{air} = 15\ mm^2/s, R = 287\ J/kgk, \tau_{0_{mid}}, D$

$$\rho = \frac{P}{RT} = \frac{100 \times 10^3}{252 \times 293} = 1.189\ kg/m^3$$

At mid of plate

$$C_f = \frac{0.059}{(Re_x)^{\frac{1}{5}}} = \frac{0.059}{\left(\frac{U_\infty x}{\upsilon}\right)^{\frac{1}{5}}} = \frac{0.059}{\left(\frac{U_\infty \times 0.5}{15 \times 10^{-6}}\right)^{\frac{1}{5}}}$$

$$Re = \frac{U_\infty x}{\upsilon} = 10^6 = \frac{U_\infty \times 0.5}{15 \times 10^{-6}} \quad \Rightarrow \quad U_\infty = 30\ m/sec$$

$$\therefore \quad C_f = 3.723 \times 10^{-3}$$

$$\tau_{0_{mid}} = \frac{1}{2}\rho U_\infty^2 C_F = 1.99 \; N/m^2$$

$$Re_L = \frac{U_\infty L}{\upsilon} = \frac{30 \times 1}{15 \times 10^{-6}} = 2 \times 10^6$$

$$\therefore \quad C_D = \frac{0.072}{(Re_L)^{\frac{1}{5}}} = 3.9548 \times 10^{-3}$$

$$D_{both\ sides} = \frac{1}{2}\rho U_\infty^2 C_D \; L \times B \times 2$$

$= 1.269\ N$

Q.40). Following the procedure leading to the Blasius equation:

$ff'' + 2f''' = 0$ for a two dimensional laminar layer on a flat plate, show that the pressure gradient a cross such a boundary layer is $\frac{\partial P}{\partial y} = \frac{\rho U_\infty^2}{2x}\sqrt{\frac{\upsilon}{U_\infty x}}\, F(f, f', f'')$,

where x = distance along the plate,
y = boundary layer transverse coordinate, U_∞ = free stream velocity, ρ = fluid density, υ = fluid kinematic viscosity, f = Blasius similarity function.
Carry out order of magnitude analysis:

$$\frac{1}{\rho}\frac{\partial P}{\partial y} = \upsilon\frac{\partial^2 \mathrm{v}}{\partial y^2} - \left(u\frac{\partial \mathrm{v}}{\partial x} + \mathrm{v}\frac{\partial \mathrm{v}}{\partial y}\right)$$

For stream function $\varphi(x, y)$: $u = \frac{\partial \varphi}{\partial y}, \mathrm{v} = -\frac{\partial \varphi}{\partial x}$

$$\frac{1}{\rho}\frac{\partial P}{\partial y} = -\upsilon\frac{\partial^3 \varphi}{\partial x \partial y^2} + \frac{\partial \varphi}{\partial y}.\frac{\partial^2 \varphi}{\partial x^2} - \frac{\partial \varphi}{\partial x}\frac{\partial^2 \varphi}{\partial x \partial y} \quad \mathbf{(1)}$$

Blasius procedure:

$$\delta = \sqrt{\frac{\upsilon x}{U_\infty}} = \frac{x}{\sqrt{Re_x}} \text{ measure of B. L. absolute thickness}$$

Define $\varphi = U_\infty\, \delta\, f(\eta)$

Where $\eta = y/\delta$, $f(\eta)$ = universal similarity function

$$\varphi = \sqrt{\upsilon U_\infty x}\, f \,, \eta = y\sqrt{\frac{U_\infty}{\upsilon x}}$$

$$\frac{\partial \varphi}{\partial x} = \frac{\partial}{\partial x}\sqrt{\upsilon U_\infty x}\, f + \sqrt{\upsilon U_\infty x}\;\frac{\partial f}{\partial \eta}.\frac{\partial \eta}{\partial x}$$

$$\frac{\partial f}{\partial \eta} = f', \qquad \frac{\partial \eta}{\partial x} = y\sqrt{\frac{U_\infty}{\upsilon}} \times \frac{d}{dx}\left(x^{-\frac{1}{2}}\right) = -\frac{y}{2}\sqrt{\frac{U_\infty}{\upsilon}}\, x^{-\frac{3}{2}} = -\frac{\eta}{2x}$$

$$\therefore \frac{\partial \varphi}{\partial x} = \frac{1}{2}\sqrt{\frac{\upsilon U_\infty}{x}}\, f - \frac{1}{2}\sqrt{\frac{\upsilon U_\infty}{x}}\, f'\eta = \frac{1}{2}\sqrt{\frac{\upsilon U_\infty}{x}}\,(f - \eta f');$$

$$\frac{\partial \varphi}{\partial y} = \sqrt{\upsilon U_\infty x}\, f'\frac{\partial \eta}{\partial y} = \sqrt{\upsilon U_\infty x}\sqrt{\frac{U_\infty}{\upsilon x}}\, f' = U_\infty f' \,,$$

$$\frac{\partial^2 \varphi}{\partial x^2} = \frac{1}{2} \times -\frac{1}{2x}\sqrt{\frac{\upsilon U_\infty}{x}}\,(f - \eta f') + \frac{1}{2}\sqrt{\frac{\upsilon U_\infty}{x}}\left(f'.\frac{\partial \eta}{\partial x} - \frac{\partial \eta}{\partial x}f' - \eta\frac{\partial f'}{\partial \eta}.\frac{\partial \eta}{\partial x}\right)$$

$$\frac{\partial^2 \varphi}{\partial x^2} = -\frac{1}{4x}\sqrt{\frac{\upsilon U_\infty}{x}}\,(f - \eta f') + \frac{1}{2}\sqrt{\frac{\upsilon U_\infty}{x}} \times \left(-\eta f''\frac{\partial \eta}{\partial x}\right)$$

$$= -\frac{1}{4x}\sqrt{\frac{\upsilon U_\infty}{x}}\,(f - \eta f') + \frac{1}{2}\sqrt{\frac{\upsilon U_\infty}{x}}\left(-\eta f''\left(-\frac{\eta}{2x}\right)\right)$$

$$= -\frac{1}{4x}\sqrt{\frac{\upsilon U_\infty}{x}}\,(f - \eta f' - \eta^2 f'')$$

$$\frac{\partial^2\varphi}{\partial x\partial y} = \frac{\partial^2}{\partial x\partial y}\left(\sqrt{\upsilon U_\infty x}\, f\right) = \frac{\partial}{\partial x}\cdot\frac{\partial\varphi}{\partial y} = \frac{\partial}{\partial x}(U_\infty f')$$

$$= U_\infty\frac{\partial f'}{\partial\eta}\cdot\frac{\partial\eta}{\partial y} = -\frac{U_\infty}{2x}\eta f''$$

$$= \frac{\partial^3\varphi}{\partial x\partial y^2} = \frac{\partial}{\partial x}\,\frac{\partial^2\varphi}{\partial y^2} = \frac{\partial}{\partial x}\left(\frac{\partial}{\partial y}\,U_\infty f'\right)$$

$$= U_\infty\frac{\partial}{\partial x}\left(\frac{\partial f'}{\partial\eta}\cdot\frac{\partial\eta}{\partial y}\right) = U_\infty\frac{\partial}{\partial x}\left(f''\sqrt{\frac{U_\infty}{\upsilon x}}\right)$$

$$= U_\infty\times\left[f''\times -\frac{1}{2x}\sqrt{\frac{U_\infty}{\upsilon x}} + \sqrt{\frac{U_\infty}{\upsilon x}}\,\frac{\partial f''}{\partial\eta}\cdot\frac{\partial\eta}{\partial y}\right]$$

$$= -\frac{U_\infty}{2x}\sqrt{\frac{U_\infty}{\upsilon x}}\,(f'' + \eta f''')$$

Now sub in (1)

$$\frac{1}{\rho}\frac{\partial P}{\partial y} = \upsilon\frac{U_\infty}{2x}\sqrt{\frac{U_\infty}{\upsilon x}}\,(f'' + \eta f''') + U_\infty f'\left(-\frac{1}{4x}\sqrt{\frac{\upsilon U_\infty}{x}}\,(f - \eta f' - \eta^2 f'')\right)$$

$$-\frac{1}{2}\sqrt{\frac{\upsilon U_\infty}{x}}\,(f - \eta f')\times\left(-\frac{U_\infty}{2x}\,\eta f''\right).$$

$$= \frac{\upsilon U_\infty}{4x}\sqrt{\frac{U_\infty}{\upsilon x}}\left[2f'' + 2\,\eta f'' - f\,f' + \eta f'^2 + \eta^2 f' f'' + f\eta f'' - \eta^2 f' f''\right]$$

$$= \frac{\upsilon U_\infty}{4x}\sqrt{\frac{U_\infty}{\upsilon x}}\left[\eta(\underline{2\,f''' + ff''}) + 2f'' - ff' + \eta f'^2\right]$$

Zero (Blasius equ)

$$= \frac{\upsilon U_\infty}{4x}\sqrt{\frac{U_\infty}{\upsilon x}}\left(2f'' - ff' + \eta f'^2\right)$$

$$\frac{\partial P}{\partial y} = \frac{\partial P}{\partial \eta}\cdot\frac{\partial \eta}{\partial y} \qquad \frac{\partial \eta}{\partial y} = \sqrt{\frac{U_\infty}{\upsilon x}}$$

$$\frac{\partial P}{\partial \eta} = \rho\frac{\upsilon U_\infty}{4x}\sqrt{\frac{U_\infty}{\upsilon x}}\left(2f'' - ff' + \eta f'^2\right) \times \sqrt{\frac{\upsilon x}{U_\infty}}$$

$$= \rho\frac{\upsilon U_\infty}{4x}\left(2f'' - ff' + \eta f'^2\right)$$

Q.41). Suppose that the shear stress across a two dimensional laminar, boundary layer with zero pressure gradient can be expressed as

$$\tau = \tau_0/\cosh^2\eta$$

Where τ_0 = wall shear stress, $\eta = y/\delta$ = nondimension transverse coordinate : δ = boundary layer absolute thickness. For such a boundary layer establish:

i) δ *ii)* δ^* *iii)* τ *iv)* C_P wall friction coefficient each in terms of the Reynolds number $(U_\infty x/\upsilon)$; where U_∞ = external velocity, x = distance from the L.E indicate the inherent weaknesses of the above shear stress assumption.

N.B $d(\tanh x)/dx = \cosh^2 x$; in integrations concerning δ^* & θ substitution of $t = \tanh y$ may prove helpful.

$$\tau = \tau_0/\cosh^2 \eta$$

$$\tau = \mu \frac{du}{dy} = \frac{\tau_0}{\cosh^2 \eta} \qquad \frac{du}{dy} = \frac{du}{d\eta} \cdot \frac{d\eta}{dy} = \frac{1}{\delta} \frac{du}{d\eta}$$

$$\frac{\mu}{\delta} \frac{du}{d\eta} = \frac{\tau_0}{\cosh^2 \eta} \quad \Rightarrow \quad du = \frac{\tau_0 \,.\, \delta}{\mu} \cdot \frac{d\eta}{\cosh^2 \eta}$$

$$u = \frac{\tau_0 \,.\, \delta}{\mu} \tanh \eta + C$$

$$\eta \to \infty \qquad \tanh \eta = 1 \qquad u = U_\infty \qquad U_\infty = \frac{\tau_0 \,.\, \delta}{\mu}$$

$$\eta \to 0 \qquad u = 0 \qquad \tanh \eta = 0 \qquad C = 0$$

$$\therefore \frac{u}{U_\infty} = \tanh \eta$$

Assume $\tanh \eta = t$

$$dt = \frac{d\eta}{\cosh^2 \eta} = \frac{\cosh^2 \eta - \sinh^2 \eta}{\cosh^2 \eta} d\eta$$
$$= (1 - \tanh^2 \eta)$$

$$\therefore d\eta = \frac{dt}{1 - t^2}$$

$$\delta^* = \delta \int_0^1 (1 - \tanh \eta) \, d\eta$$

$$\delta^* = \delta \int_0^1 \frac{1 - t}{1 - t^2} \, dt$$

$$= \delta \int_0^1 \frac{dt}{1+t} = \delta \ln(1+t)]_0^1 = \delta \ln 2 = 0.693\,\delta$$

$$\theta = \delta \int_0^1 f(\eta)\,(1 - f(\eta))\; d\eta$$

$$= \delta \int_0^1 \tanh\eta\,(1 - \tanh\eta)\; d\eta$$

$$= \delta \int_0^1 \frac{t(1-t)}{1-t^2}\, dt = \delta \int_0^1 \frac{t}{1+t}\, dt$$

$$= \delta[t - \ln(t+1)]_0^1 = \delta[1 - \ln(2)] = 0.307\,\delta = \alpha\,\delta$$

$$\tau_0 = \rho U_\infty^2 \frac{\partial\delta}{\partial x}\,\alpha$$

$$\frac{\mu U_\infty}{\delta} = \rho U_\infty^2 \frac{\partial\delta}{\partial x} \times 0.307$$

$$0.307\,\delta\,\partial\delta = \frac{\upsilon}{U_\infty} dx$$

$$\frac{\delta^2}{2} = 3.257 \frac{\upsilon}{U_\infty} x + C \qquad \text{at } x = 0 \quad \delta = 0 \Rightarrow C = 0$$

$$\delta = \frac{2.55\,x}{\sqrt{Re_x}}$$

$$\delta^* = 0.693\,\delta = \frac{1.769\,x}{\sqrt{Re_x}}$$

$$\theta = 0.307\,\delta = \frac{0.78\,x}{\sqrt{Re_x}}$$

$$\frac{C_f}{2} = \frac{\tau_0}{\rho U_\infty^2}$$

$$C_f = \frac{2\,\mu U_\infty}{\delta\,\rho U_\infty^2} = \frac{2\upsilon}{\delta U_\infty} \cdot \frac{x}{x} = \frac{0.78}{\sqrt{Re_x}}$$

Q.42). One method B.L. control is to use suction as shown in the sketch. Clearly this helps to limits B.L. growth and delay separation. Consider the suction velocity v_s at the plate surface to be uniform. Assume $v_s/U_\infty \ll 1$.

For steady T.B.L. over flat plate show that

a) At the laminar sub – layer

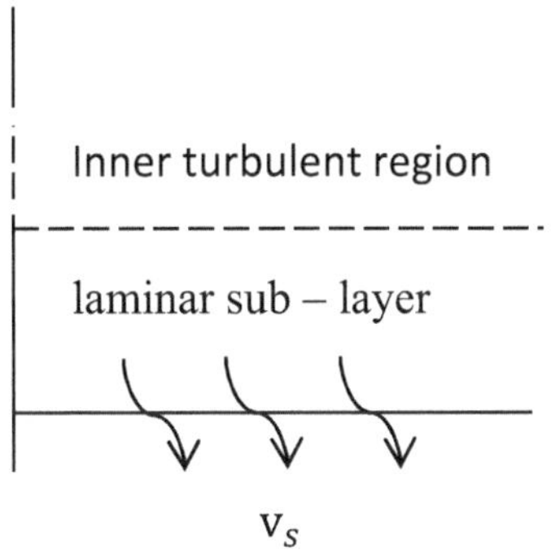

$$\frac{\bar{u}}{u^*} = \frac{u^*}{v_s}\left[1 - exp\left(-\frac{v_s\,y}{\upsilon}\right)\right]$$

b) At inner turbulent region

$$-\frac{2u^*}{v_s}\left[1 - \frac{v_s\,\bar{u}}{u^{*2}}\right] = \frac{1}{k}\ln\left(\frac{yu^*}{\upsilon}\right) + c$$

Note. $\tau = \mu\frac{\partial \bar{u}}{\partial y} - \rho u'v'$

At $y = 0$ $u = 0$ $v = -v_s$

Near the porous $\frac{\partial \bar{u}}{\partial x} = 0$

Solution:

Since $\frac{\partial \bar{u}}{\partial x} = 0$ near the porous boundary

$\therefore$ from continuity equation $\frac{\partial \bar{v}}{\partial y} = 0$

Integration then gives $\bar{v} = constant$ or at $y = 0,\ \bar{v} = -v_s$

Hence from the momentum equation by neglecting the pressure variation

$$-\rho v_s \frac{\partial \bar{u}}{\partial y} = \frac{\partial \tau}{\partial y} \quad (1)$$

Then integration both side with respect to y we get

$$\tau = -\rho v_s \bar{u} + f(x)$$

Now $u = 0$ and $\tau = \tau_w$ when $y = 0$

$$\therefore \tau = \tau_w - \rho v_s \bar{u} \quad (2)$$

$$i.e.\ \ \mu \frac{\partial \bar{u}}{\partial y} - \rho \overline{u'v'} = \tau_w - \rho v_s \bar{u} \quad (3)$$

Earlier work suggests that this flow near the boundary consists of
1- A Laminar Sub – Layer. Viscous stress≫ Reynolds stress.

Then equation (3) reduced to

$\mu \frac{\partial \bar{u}}{\partial y} = \tau_w - \rho v_s \bar{u}$ by neglecting Reynolds stress term.

and $u^* = \sqrt{\frac{\tau_w}{\rho}}$

then $\upsilon \cdot \frac{\partial \bar{u}}{\partial y} = u^{*2} - \mathrm{v}_s \bar{u}$ **(4)**

separating the variables $\bar{u}$ and y gives

$$\frac{\upsilon\, \partial \bar{u}}{u^{*2} - \mathrm{v}_s \bar{u}} = dy \quad \text{integrate both side}$$

We get

$$-\frac{\upsilon}{\mathrm{v}_s} \ln\left(u^{*2} - \mathrm{v}_s \bar{u}\right) = y + constant$$

When $y = 0, \quad \bar{u} = 0 \qquad \therefore const. = -\frac{\upsilon}{\mathrm{v}_s} \ln\left(u^{*2}\right)$

$$\therefore \frac{\mathrm{v}_s}{\upsilon} = -\ln\left[1 - \frac{\mathrm{v}_s \bar{u}}{u^{*2}}\right]$$

$$\text{or } 1 - \frac{\mathrm{v}_s \bar{u}}{u^{*2}} = e^{-\frac{\mathrm{v}_s y}{\upsilon}}$$

$$\frac{\mathrm{v}_s \bar{u}}{u^{*2}} = 1 - e^{-\frac{\mathrm{v}_s y}{\upsilon}}$$

or

$$\frac{\bar{u}}{u^*} = \frac{u^*}{\mathrm{v}_s}\left[1 - exp\left(-\frac{\mathrm{v}_s y}{\upsilon}\right)\right] \tag{5}$$

2- At inner turbulent region in which the viscous stress can be neglected, when compared with Reynolds stress.

$\therefore$ equation (3) gives

$$-\rho u'\mathrm{v}' = \tau_w - \rho \mathrm{v}_s \bar{u} \tag{6}$$

In the present flow $\frac{\partial \bar{u}}{\partial y} > 0$

$\therefore \left|\frac{\partial \bar{u}}{\partial y}\right| = \frac{\partial \bar{u}}{\partial y}$ and so Prandtl's mixing

Length hypothesis gives:

$$-\rho\overline{u'v'} = \rho\ell^2\left(\frac{\partial \bar{u}}{\partial y}\right)^2, \quad \text{put} \quad \ell = ky$$

Where $k = constant$

$$\therefore -\rho k^2 y^2\left(\frac{\partial \bar{u}}{\partial y}\right)^2 = \tau_w - \rho v_s \bar{u}$$

or

$$\left(ky\left(\frac{\partial \bar{u}}{\partial y}\right)\right)^2 = u^{*2} - v_s \bar{u} \tag{7}$$

Take the square root and separate variable we get

$$\frac{k\partial \bar{u}}{\left[u^{*2} - v_s \bar{u}\right]^{1/2}} = \frac{\partial y}{y}$$

Integrate both side gives

$$-\frac{2k}{v_s}\left[u^{*2} - v_s \bar{u}\right]^{\frac{1}{2}} = \ln y + c \tag{8}$$

Re- arranging equation (8) and make

$\ln y \rightarrow \ln\left(\frac{yu^*}{\upsilon}\right)$ to get dimensionless

$\therefore$ equation (8) becomes

$$-\frac{2u^*}{v_s}\left(1-\frac{v_s\bar{u}}{u^{*2}}\right)^{\frac{1}{2}}=\frac{1}{k}\ln\left(\frac{yu^*}{\upsilon}\right)+c \tag{9}$$

By letting $v_s \to 0$ in the above eq.'s the earlier " no suction " eq.'s are obtained.

Q.43). To illustrate the behavior of the B.L. solution for unsteady at stagnation point of boundary layer flow, let $U_{(x)}$ is proportional to $a_{(t)}$ which is an arbitrary constant, a function of time. Where

$$\varphi=\sqrt{\upsilon x U_{(x)}}\,f(\eta) \qquad \eta=y\sqrt{\frac{U_{(x)}}{\upsilon x}}$$

And full navier – stokes equation in x – direction

$$\frac{\partial u}{\partial t}+u\frac{\partial u}{\partial x}+v\frac{\partial u}{\partial y}=-\frac{1}{\rho}\frac{\partial P}{\partial x}+\upsilon\frac{\partial^2 u}{\partial y^2}$$

Show that the solution as follows.

$$\frac{\dot{a}}{a^2}\left[f'(\eta)+\frac{\eta}{2}f''(\eta)\right]+f'^2(\eta)-f(\eta)f''(\eta)=x\left(\frac{\dot{a}}{a^2{}_{(t)}}+1\right)+f'''(\eta)$$

Solution:

- Now a similarity solution can be found even for the unsteady stagnation point boundary layer flow:

Now:

- The boundary layer equation is:

$$\frac{\partial u}{\partial x}+\frac{\partial v}{\partial y}=0$$

$$\frac{\partial u}{\partial t}+u\frac{\partial u}{\partial x}+v\frac{\partial u}{\partial y}=-\frac{1}{\rho}\frac{\partial P}{\partial x}+\upsilon\frac{\partial^2 u}{\partial y^2}$$

- Let $U = a(t)x$ **(1)**

- The stream function is:

$$\varphi = \sqrt{\upsilon x U}\, f(\eta)$$

$$= \sqrt{\upsilon x^2 a(t)}\, f(\eta)$$

$$\varphi = \sqrt{\upsilon\, a(t)}\, x\, f(\eta) \qquad \textbf{(2)}$$

$$\eta = y\sqrt{\frac{U}{\upsilon x}} = y\sqrt{\frac{a(t)x}{\upsilon x}}$$

$$\eta = y\sqrt{\frac{a(t)}{\upsilon}} \qquad \textbf{(3)}$$

- Now, the stream function equation is:

$$u = \frac{\partial\varphi}{\partial y} \qquad \mathrm{v} = -\frac{\partial\varphi}{\partial x}$$

So that:

$$u = \frac{\partial\varphi}{\partial y} = \frac{\partial\varphi}{\partial\eta}\frac{\partial\eta}{\partial y}$$

$$= \sqrt{\upsilon\, a(t)}\, f'(\eta)\cdot\sqrt{\frac{a(t)}{\upsilon}}$$

$$u = \sqrt{\frac{\upsilon\, a(t)\cdot a(t)}{\upsilon}}\, f'(\eta)$$

$$u = a(t)f'(\eta) \qquad \textbf{(4)}$$

$$\mathrm{v} = -\frac{\partial\varphi}{\partial x} = -\frac{\partial\varphi}{\partial\eta}\cdot\frac{\partial\eta}{\partial x}$$

$$\mathrm{v} = -f(\eta) \times \sqrt{\upsilon\, a(t)} + 0 \tag{5}$$

- Now, the partial differential equation is:

$$* \frac{\partial u}{\partial y} = \frac{\partial}{\partial y}\big(a(t) f(\eta)\big) = a(t)\sqrt{\frac{a(t)}{\upsilon}} f''(\eta) \tag{6}$$

$$* \frac{\partial^2 u}{\partial y^2} = a(t)\sqrt{\frac{a(t)}{\upsilon}}\sqrt{\frac{a(t)}{\upsilon}} f'''(\eta) = \frac{a_{(t)}^2}{\upsilon} f'''(\eta) \tag{7}$$

$$* \frac{\partial u}{\partial x} = \frac{\partial}{\partial x}\big(a(t) f'(\eta)\big) = a(t) f'(\eta) \tag{8}$$

$$* \frac{\partial u}{\partial t} = f'(\eta)\dot{a}(t) + a(t) \times y\sqrt{\frac{1}{\upsilon}} \times f''(\eta) \times \frac{1}{2} a(t)^{-\frac{1}{2}}\, \dot{a}$$

$$(t) = \dot{a}(t)\left[f'(\eta) + \frac{y}{2}\sqrt{\frac{a(t)}{\upsilon}} f''(\eta)\right] = \dot{a}(t)\left[f'(\eta) + \frac{\eta}{2} f''(\eta)\right] \tag{9}$$

$$\therefore -\frac{1}{\rho}\frac{\partial P}{\partial x} = \frac{\partial U}{\partial t} + U\frac{\partial U}{\partial x} \qquad U = a(t)x$$

$$= \dot{a}_{(t)} + a_{(t)}x \cdot a_{(t)}$$

$$= \dot{a}_{(t)}x + a_{(t)}^2 x$$

$$\therefore -\frac{1}{\rho}\frac{\partial P}{\partial x} = a_{(t)}^2 x\left[\frac{\dot{a}}{a_{(t)}^2} + 1\right] \tag{10}$$

$$\therefore \frac{\partial u}{\partial t} + u\frac{\partial u}{\partial x} + \mathrm{v}\frac{\partial u}{\partial y} = -\frac{1}{\rho}\frac{\partial P}{\partial x} + \upsilon\frac{\partial^2 u}{\partial y^2}$$

$$\therefore \dot{a}(t)\left[f'(\eta) + \frac{\eta}{2} f''(\eta)\right] + a(t)f'(\eta) \times a(t)f'(\eta) - f(\eta)\sqrt{\upsilon\, a(t)}$$

$$\times a(t)\sqrt{\frac{a(t)}{\upsilon}}f''(\eta)$$

$$= a^2x\left(\frac{\dot{a}}{a^2}+1\right)+\upsilon\frac{a^2}{\upsilon}f'''(\eta) \qquad\qquad \div a^2$$

$$\therefore \frac{\dot{a}}{a^2}\left[f'(\eta)+\frac{\eta}{2}f''(\eta)\right]+f'^2(\eta)-f(\eta)f''(\eta)$$

$$= x\left(\frac{\dot{a}}{a^2}+1\right)+f'''(\eta) \tag{11}$$

At stagnation point $P \to 0$ then the term

$$x\left(\frac{\dot{a(t)}}{a^2_{(t)}}+1\right)\to 0.$$

Q.44). To illustrate the behavior of the boundary – layer solution let U is proportional to x^m. The irrotational flow corresponds to this is the flow past a wedge of angle radians with a complex potential and velocity given by

$$w=\phi+i\varphi=\frac{cz^{m+1}}{m+1}=\frac{cr^{m+1}(\cos(m+1)\theta+i\sin(m+1)\theta)}{m+1} \tag{1}$$

$$\frac{\partial w}{\partial z}=u-iv=cr^m(\cos m\theta\ +i\sin m\theta) \tag{2}$$

For $(\theta=0)$ on the surface of the wedge gives $U=cr^m$

For the similarity form of the solution use

$$\varphi=\sqrt{\upsilon\, U(x)x}\, f(\eta) \qquad \text{with} \qquad \eta=\sqrt{\frac{U(x)}{x\,\upsilon}}.$$

Show that Falkner – skan equation for flow past a wedge as follows:

$$f'''+0.5(m+1)ff''+m[1-(f')^2]=0$$

Solution:

- We consider symmetric wedge as shown in figure. We shall first deal with the outer invisiced potential flow; whose velocity distribution leads to asymptotic boundary conditions for the inner flow calculation.

- In the boundary layer coordinate where we measure (x) along the upper surface of the body and (y) perpendicular to it, we therefore obtain exactly the power law distribution:

Now, - As shown by Euler's equation:

$$-\frac{1}{\rho}\frac{\partial p}{\partial x} = mc^2x^{2m-1} \tag{1}$$

- From the power law distribution:

$$U(x) = cx^m \tag{2}$$

- The solution should be in the form:

$$\varphi = \sqrt{\upsilon\, U(x)x}\, f(\eta)$$

$$\eta = y\sqrt{\frac{U(x)}{\upsilon x}}$$

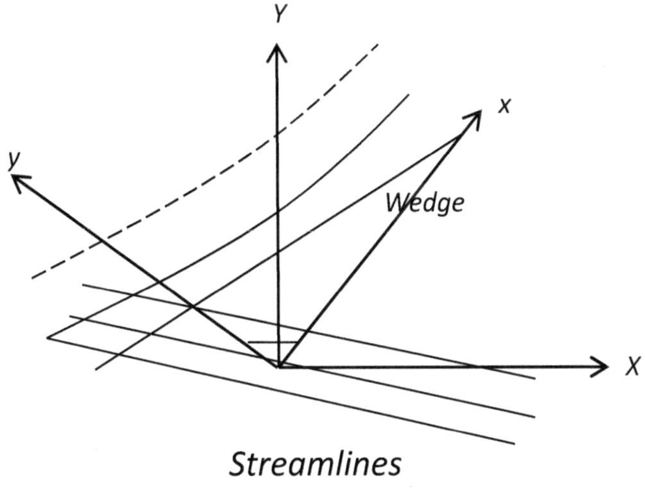

- the boundary layer equation are:

$$\frac{\partial u}{\partial x}+\frac{\partial \mathrm{v}}{\partial y}=0$$

$$u\frac{\partial u}{\partial x}+\mathrm{v}\frac{\partial u}{\partial y}=-\frac{1}{\rho}\frac{\partial P}{\partial x}+\upsilon\frac{\partial^2 u}{\partial y^2}$$

- the stream function equation is:

$$u\frac{\partial \varphi}{\partial y} \qquad \mathrm{v}=-\frac{\partial \varphi}{\partial x}$$

Now:

$$\varphi=\sqrt{\upsilon\, U(x)x}\, f(\eta) \qquad U(x)=cx^m$$

$$\varphi=\sqrt{\upsilon\, cx^m\cdot x}\, f(\eta)$$

$$\varphi=\sqrt{\upsilon\, cx^{m+1}}\, f(\eta) \qquad \textbf{(3)}$$

$$\eta=y\sqrt{\frac{cx^m}{\upsilon x}}$$

$$\eta=y\sqrt{\frac{c}{\upsilon}x^{m-1}} \qquad \textbf{(4)}$$

$$\therefore u=\frac{\partial \varphi}{\partial y}=\frac{\partial \varphi}{\partial \eta}\frac{\partial \eta}{\partial y}=f'(\eta)\times\sqrt{\upsilon\, cx^{m+1}}\times\sqrt{\frac{c}{\upsilon}x^{m-1}}$$

$$=f'(\eta)\times\sqrt{\frac{\upsilon\, c^2x^{m+1+m-1}}{\upsilon}}=f'(\eta)\, cx^m$$

$$u=U(x)f'(\eta) \qquad \textbf{(5)}$$

$$\therefore \mathrm{v}=-\frac{\partial \varphi}{\partial x}=-\frac{\partial \varphi}{\partial \eta}\times\frac{\partial \eta}{\partial x}$$

$$v = -\left[\sqrt{\upsilon c x^{m+1}} f'(\eta)\frac{\partial \eta}{\partial x} + f(\eta)\sqrt{\upsilon\, c} \times \frac{m+1}{2} x^{\frac{m+1}{2}-1}\right]$$

$$= -\left[\sqrt{\upsilon\, c x^{m+1}} f'(\eta) \times y\sqrt{\frac{c}{\upsilon}} \times \frac{m-1}{2} x^{\frac{m-1}{2}-1} + f(\eta)\sqrt{\upsilon\, c}\ \frac{m+1}{2} x^{\frac{m-1}{2}}\right]$$

$$= -\left[\sqrt{\upsilon\, c x^{m+1}}\ f'(\eta)\ \eta\sqrt{\frac{\upsilon}{c x^{m-1}}}\ \frac{m-1}{2} x^{\frac{m-1}{2}-1}\sqrt{\frac{c}{\upsilon}} + \sqrt{\upsilon c} f(\eta)\frac{m+1}{2} x^{\frac{m-1}{2}}\right]$$

$$= -\left[\sqrt{\upsilon\, c x^{m+1}}\ f'(\eta)\ \eta \times \frac{m-1}{2} \times \frac{x^{\frac{m-1}{2}} x^{-1}}{x^{\frac{m-1}{2}}} + \sqrt{\upsilon c} f(\eta)\frac{m+1}{2} x^{\frac{m-1}{2}}\right]$$

$$= -\left[\sqrt{\upsilon\, c x^{m+1}}\ \frac{m-1}{2} \times \frac{1}{x} f'(\eta)\ \eta + \sqrt{\upsilon c} f(\eta)\frac{m+1}{2} x^{\frac{m-1}{2}}\right]$$

$$= -\left[\sqrt{\upsilon\, c\frac{x^{m+1}}{x^2}}\ \frac{m-1}{2} f'(\eta)\ \eta + \sqrt{\upsilon c}\ \frac{m+1}{2} f(\eta) x^{\frac{m-1}{2}}\right]$$

$$v = -\left[\sqrt{\upsilon\, c\, x^{m-1}}\ \frac{m-1}{2} f'(\eta)\ \eta + \sqrt{\upsilon\, c\, x^{m-1}}\ \frac{m+1}{2} f(\eta)\right] \quad \textbf{(6)}$$

Now:

$$\frac{\partial u}{\partial x} = ? \qquad \frac{\partial u}{\partial y} = ? \qquad \frac{\partial^2 u}{\partial y^2} = ?$$

So:

$$\frac{\partial u}{\partial x} = \frac{\partial}{\partial x}\big(U(x) f'(\eta)\big) = \frac{\partial}{\partial x}\big(c x^m f'(\eta)\big)$$

$$= c x^m\left(\frac{m-1}{2} y\sqrt{\frac{c}{\upsilon}} x^{\frac{m-1}{2}-1}\ f''(\eta) + f'(\eta)\ m\ c x^{m-1}\right)$$

$$\therefore \frac{\partial u}{\partial x} = U(x)\left[y\sqrt{\frac{c}{\upsilon}} x^{\frac{m-1}{2}} \cdot x^{-1} f''(\eta)\frac{m-1}{2}\right] + f'(\eta)\ m\ x^{-1} c x^m$$

$$= U(x)\left[y\sqrt{\frac{c}{\upsilon}x^{m-1}}\,x^{-1}\frac{m-1}{2}f''(\eta)\right] + U(x)\,x^{-1}\,m\,f'(\eta)$$

$$\frac{\partial u}{\partial x} = U(x)x^{-1}\left[\eta f''(\eta)\frac{m-1}{2}\;f'(\eta)m\right] \quad \textbf{(7)}$$

So:

$$\frac{\partial u}{\partial y} = \frac{\partial}{\partial y}\left(U(x)f'(\eta)\right) = U(x)\sqrt{\frac{c}{\upsilon}x^{m-1}}\,f''(\eta)$$

$$= cx^m\sqrt{\frac{c}{\upsilon}x^{m-1}}\,f''(\eta) \quad \textbf{(8)}$$

Now:

$$\frac{\partial^2 u}{\partial y^2} = \frac{\partial}{\partial y}\left[U(x)\sqrt{\frac{c}{\upsilon}x^{m-1}}\,f''(\eta)\right]$$

$$= U(x)\sqrt{\frac{c\,x^{m-1}}{\upsilon}}\sqrt{\frac{c\,x^{m-1}}{\upsilon}}\;f'''(\eta)$$

$$= cx^m\cdot\frac{c\,x^{m-1}}{\upsilon}f'''(\eta) \quad \textbf{(9)}$$

$$\therefore\; u\frac{\partial u}{\partial x} + \mathrm{v}\frac{\partial u}{\partial y} = -\frac{1}{\rho}\frac{\partial P}{\partial x} + \upsilon\frac{\partial^2 u}{\partial y^2}$$

$$cx^m f'(\eta)\left[cm\,x^{m-1}\left(\eta f''(\eta)\frac{m-1}{2} + f'(\eta)m\right)\right]$$
$$-\left[\sqrt{\upsilon\,cx^{m-1}}\frac{m-1}{2}f'(\eta)\eta + \sqrt{\upsilon\,cx^{m-1}}\frac{m+1}{2}f(\eta)\right]$$
$$\times\, cx^m\sqrt{\frac{c}{\upsilon}x^{m-1}}\;f''(\eta) = mc^2x^{2m-1} + \upsilon\,cx^m\frac{c\,x^{m-1}}{\upsilon}f'''(\eta)$$

$$c^2x^{2m-1}\left[\eta f'f''\frac{m-1}{2} + m\,f'^2\right] - \left[c^2x^{2m-1}\left(\frac{m-1}{2}\eta f'f'' + \frac{m+1}{2}ff''\right)\right]$$

$$= m\,c^2x^{2m-1} + c^2x^{2m-1}f'''$$

$$\eta f'f''\frac{m-1}{2}+mf'^2-\frac{m-1}{2}\eta f'f''-\frac{m+1}{2}ff''=m+f'''$$

$$f'''-m\,f'^2+m+\frac{m+1}{2}ff''=0$$

$$\therefore f'''+\tfrac{m+1}{2}ff''+m\left(1-f'^2\right)=0 \qquad \textbf{(10)}$$

- The solution of this so – called Falkner – Skan equation.

- The first solution of equation (10) was carried out by Blasius for a semi – infinite flat plate, where $m = 0$. Then equs (10) reduce to

$$f'''+0.5ff''=0$$

Q.45). Near the stagnation on a cylinder in a uniform flow, U_∞ the local free stream velocity is approximately $u=\frac{2U_\infty x}{R}$, thus for laminar flow, the B.L. equation near the stagnation point is

$$u\frac{\partial u}{\partial x}+\mathrm{v}\frac{\partial u}{\partial y}=4\,\frac{U_\infty^2}{R^2}x+\upsilon\frac{\partial^2 u}{\partial y^2}$$

Using the transformation

$$\eta=\frac{Ay}{x^n},\qquad \varphi=Bx^m f(\eta)$$

The equation becomes

$$\frac{d^3f}{d\eta^3}+f\frac{d^2f}{d\eta^2}-\left(\frac{df}{d\eta}\right)^2+1=0$$

Solution:

Given laminar flow B.L. equation near stagnation point of cylinder as

$$u\frac{\partial u}{\partial x}+\mathrm{v}\frac{\partial u}{\partial y}=4\,\frac{U_\infty^2}{R^2}x+\upsilon\frac{\partial^2 u}{\partial y^2} \quad \textbf{(1)}$$

Given transformation

$$\eta=\frac{Ay}{x^n}=\eta(x,y) \qquad \text{(non - dimensional B. L. coordinate)}$$

$$\varphi=Bx^m f(\eta)=\varphi(\eta,x) \qquad \text{(stream function)}$$

Need

$$u,\mathrm{v},\frac{\partial u}{\partial x},\frac{\partial u}{\partial y},\frac{\partial^2 u}{\partial y^2} \qquad \text{for eq. (1)}$$

$$u=\frac{\partial\varphi}{\partial y}=\frac{\partial\varphi}{\partial\eta}\frac{\partial\eta}{\partial y}+\frac{\partial\varphi}{\partial x}\frac{\partial x}{\partial y} \qquad \frac{\partial x}{\partial y}\to 0$$

$$\therefore u=Bx^m f'\cdot\frac{A}{x^n}$$

$$\text{or } u=ABx^{m-n}f' \quad \textbf{(2)}$$

$$\mathrm{v}=-\frac{\partial\varphi}{\partial x}=-\left[\frac{\partial\varphi}{\partial\eta}\frac{\partial\eta}{\partial x}+\frac{\partial\varphi}{\partial x}\frac{\partial x}{\partial x}\right]$$

$$=-\left[Bx^m f'\left(-nAyx^{-(n+1)}\right)+mBx^{m-1}f\right]$$

$$\therefore \mathrm{v}=nAByx^{m-n-1}f'-mBx^{m-1}f \quad \textbf{(3)}$$

Now $u=u(x,\eta)$

$$\frac{\partial u}{\partial x}=\frac{\partial u}{\partial x}\frac{\partial x}{\partial x}+\frac{\partial u}{\partial\eta}\cdot\frac{\partial\eta}{\partial x}$$

$$=(m-n)ABx^{m-n-1}f'-nA^2Byx^{m-2n-1}f'' \quad \textbf{(4)}$$

$$u\neq u(x,\eta)$$

$$\frac{\partial u}{\partial y}=\frac{\partial u}{\partial x}\frac{\partial x}{\partial y}+\frac{\partial u}{\partial \eta}\cdot\frac{\partial \eta}{\partial x}=0+ABx^{m-n}f''\cdot Ax^{-n}$$

$$=A^2Bx^{m-2n}f'' \qquad \textbf{(5)}$$

Now

$$\frac{\partial u}{\partial y}=\frac{\partial u}{\partial y}(x,\eta)$$

$$\frac{\partial^2 u}{\partial y^2}=\frac{\partial}{\partial y}\left(\frac{\partial u}{\partial y}\right)=\frac{\partial}{\partial x}\left(\frac{\partial u}{\partial y}\right)\cdot\frac{\partial x}{\partial y}+\frac{\partial}{\partial \eta}\left(\frac{\partial u}{\partial y}\right)\cdot\frac{\partial \eta}{\partial y}$$

$$=A^2Bx^{m-2n}f'''\cdot Ax^{-n}$$

$$\frac{\partial^2 u}{\partial y^2}=A^3Bx^{m-3n}f''' \qquad \textbf{(6)}$$

Now substituting equs 2→6 in (1) gives

$$(ABx^{m-n}f')[(m-n)ABx^{m-n-1}f'-nA^2Byx^{m-2n-1}f'']$$

$$+[nAByx^{m-n-1}f'-mBx^{m-1}f]\left(A^2Bx^{m-2n}f''\right)$$

$$=4\,\frac{U_\infty^2}{R^2}x+\upsilon A^3Bx^{m-3n}f'''$$

$$(m-n)A^2B^2x^{2m-2n-1}(f')^2-nA^3B^2yx^{2m-3n-1}f'f''$$

$$+nA^3B^2yx^{2m-3n-1}f'f''-mA^2B^2x^{2m-2n-1}ff''$$

$$=4\,\frac{U_\infty^2}{R^2}x+\upsilon A^3Bx^{m-3n}f'''$$

Dividing both sides by $\frac{4U_\infty^2}{R^2}x$ to get

$$\frac{(m-n)A^2B^2x^{2m-2n-1}}{\frac{4U_\infty^2}{R^2}x}(f')^2 - \frac{mA^2B^2x^{2m-2n-1}}{\frac{4U_\infty^2}{R^2}x}ff'' = 1 + \frac{\upsilon A^3Bx^{m-3n}}{\frac{4U_\infty^2}{R^2}x}f'''$$

$$E(f')^2 - Gff'' = 1 + Hf'''$$

Rearranging as

$$Hf''' + Gff'' - E(f')^2 + 1 = 0$$

Given the transformed governing eq. a

$$\frac{d^3f}{d\eta^3} + f\frac{d^2f}{d\eta^2} - \left(\frac{df}{d\eta}\right)^2 + 1 = 0$$

Comparing both equation, we should have

$$H = 1 \qquad G = 1 \qquad E = 1$$

That is

$$\frac{(m-n)A^2B^2}{\frac{4U_\infty^2}{R^2}x}x^{2m-2n-1} = 1 \quad \textbf{(7)}$$

$$\frac{mA^2B^2}{\frac{4U_\infty^2}{R^2}x}x^{2m-2n-1} = 1 \quad \textbf{(8)}$$

$$\frac{\upsilon A^3B}{\frac{4U_\infty^2}{R^2}x}x^{m-3n} = 1 \quad \textbf{(9)}$$

Given m, n, A, U_∞, R constant and independent of x then, indices are equal for each eq.

$$\text{From} \begin{matrix}(7)\\(8)\end{matrix} \quad \left.\begin{matrix}2m-2n-1=1\\2m-2n-1=1\end{matrix}\right\} \quad \textbf{(10)}$$

$$m - 3n = 1 \quad \textbf{(11)}$$

Solving for (10) and (11) gives

$n = 0 \qquad m = 1$

Then from equation (7) using the found value m and n

$$\frac{(1-0)A^2B^2}{\frac{4U_\infty^2}{R^2}x}x^{2\times1-0-1} = 1$$

$$\frac{A^2B^2}{\frac{4U_\infty^2}{R^2}x}x = 1$$

$$\therefore A^2B^2 = \frac{4U_\infty^2}{R^2} \tag{12}$$

From eq. (9)

$$\upsilon A^3B = \frac{4U_\infty^2}{R^2} \tag{13}$$

Solving for equ's. (12,13)

$$\therefore A^2B^2 = \upsilon A^3B$$

$$B = \upsilon A = \frac{4U_\infty^2}{\upsilon R^2A^3}$$

or $A^4 = \dfrac{4U_\infty^2}{\upsilon^2R^2} \qquad \therefore A = \left(\dfrac{2U_\infty}{\upsilon R}\right)^{1/2}$

and $B = \upsilon\left(\dfrac{2U_\infty}{\upsilon R}\right)^{1/2}$

$$B = \left(\frac{2\upsilon U_\infty}{R}\right)^{1/2}$$

REFERENCES

[1] J.M. Hassan, *Ph.D. lecture 2018-2019*. University of Technology: Baghdad Iraq.

[2] H. Tennekes, and J.L. Lumley, *A first Course in Turbulence*. 7th. The MIT press: Cambridge, 1981.

[3] N. Najdat, *Ph.D. lectures*. University of Baghdad: Baghdad Iraq, 2010 – 2012.

[4] M. Frank, *White, Viscous Fluid Flow*. 2nd ed. McGraw – Hill Bak Company, 1991.

[5] M. Frank, *White, Fluid Mechanics*. 2nd ed. McGraw – Hill Bak Company, 1986.

[6] L.S. Victor, and E.B. Whily, *Fluid Mechanics*. Mc Graw Hill, 2006.

[7] A. Yunus, *Cengel and John M. Cimbala, Fluid Mechanics Fundamental and application*. 2nd ed. Mc Gaw Hill Education, 2010.

[8] Herrmann Schlichting, *Boundary Layer Theory*. McGraw – Hill Company, 1987.

[9] A.A. Townsend, *The structure of turbulent shear flow*. 2nd ed. Cambridge University Press: Cambridge, 1976.

[10] M. Hanif Chaudhry, *Applied Hydraulic Transients*. VAN NOSTRAND REINHID Company: New York, 1979.

[11] L. Victor, *Streeter and E. Benjamin Wylie, Fluid Mechanics*. 9th ed. McGraw – Hill Bak Company, 1998.

[12] W.P. Graebel, *Advanced Fluid Mechanics.*, ELSEVIER Book AID International, .

[13] C. Folas, O. Manley, R. Rosa, and R. Teman, *Navier – Stokes Equations and Turbulence*. Cambridge University Press, 2001.

[14] *E.L.Houghton and N.B. Carruthers, Aerodynamics for Engineering Students*. 3rd ed. Edward Arnold Publishers Ltd., 1982.

[15] BRADSHAW, *P., The Understanding Prediction of Turbulent Flow*. University of Manchester, 1986.

[16] S. Mittal, "Flow past rotating cylinders effect of eccentricity", *Journal of Applied Mechanics,* vol. 68, pp. 15-43, 2001.

[17] J.F. Dovglas, and R.D. Matthews, *Fluid Mechanics*. 3rd ed. Longman Group limited, 1986.

Jafar Mehdi Hassan, Riyadh S. Al-Turaihi, Salman Hussien Omran, Laith Jaafer Habeeb, Alamaslamani Ammar Fadhil Shnawa

SUBJECT INDEX

A

Acceleration 9, 13, 74, 84, 196
  buoyant 9
Air 84, 87, 88, 91, 126, 213, 240
  chamber 84, 87, 88
  compressor 87
  cushion 91
  density 126
  flow 126, 213, 240
  inlet valves 91
  pockets 91
Aircraft wings 197
Aluminum alloys 80
Angle 205, 241, 242, 273
  radians 273
Approximation 43, 53
  dimensional boundary layer 53
Asymptotic 6, 257
  Invariance 6
  air 257
Atypical velocity scale 57

B

Bernouli equation 178
Bernoulli's theorem 166
Blasius 151, 165, 166, 168, 260
  formula friction factor in pipes 165
  problem 168
  procedure 260
  similarity function 260
  solution 151, 166
Blasius equation 167, 253, 260
  celebrated nonlinear 167
Boundary layer 39, 57, 106, 176, 190, 198, 208, 220, 235, 258, 270, 275
  equation 270, 275
  flow 39, 190, 198, 270
  laminer 176
  measurements 57
  problems 106
  thickness 208, 220, 235, 258
Boundary-Layer velocity profiles 234

C

Channel flow, developed 30
Closure problem of turbulence theory 5
Combustion processes 1
Compressed air 87
Computer models solutions 151
Conditions 64, 70, 79, 82, 90, 142, 167, 181, 188, 191, 197
  emergency 90
  nonhomogeneous rock 82
  steady 64
Convective 17
  time scale 17
  velocity 17
Convergent chanal 171
Convergent channal 171
Coriolios parameter 12
Coriolis force 11

D

Decomposition velocity 24
Density 8, 48, 64, 65, 66, 69, 79, 81, 177, 197, 257
  change 66
  increasing 69
  low 197
  small 8
Derivation of reynolds stress equation of motion 47
Device and methods for controlling transients 84
Differential equations 42, 63
Diffusion 7, 8, 9, 11, 14, 16, 17, 55, 57, 59, 151, 197
  agent, effective 14

Jafar Mehdi Hassan, Riyadh S. Al-Turaihi, Salman Hussien Omran, Laith Jaafer Habeeb, Alamaslamani Ammar Fadhil Shnawa

by random motion 9
equation 7, 11
length scale 16
molecular 7, 8, 9
of turbulent 7
time scale 17
transverses viscous 151
Dimensional 6, 21, 39, 41, 178
analysis 6, 39
of vorticity 21
reasoning 41
shear flow equations 178
Direction 8, 25, 35, 36, 55, 64, 66, 72, 100, 102, 110, 135, 151, 158, 160, 195, 270
transverse 35, 36, 100
upstream 64, 66
Dissipation of TKE by molecular viscosity force 55
Downstream 63, 64, 66, 67, 72, 90, 91, 111, 116, 151, 152, 157, 191, 197
convections 151
direction 72
distance 111
moving 66
Drag 197
airfoils 197
Drag coefficient 18, 192, 194, 195, 212, 216, 218, 225, 235, 255
dimensional surface friction 194
Dynamic equation 72, 73

E

Earthquake loads 88
Eddies 18, 20, 21, 22, 32, 33, 96, 97, 107, 108, 188
effective 108
formation 32
large 18, 22, 96, 97
largest 20
smallest 20
types 21
Effect 6, 22, 28, 46
of motion 22
of wall roughness on turbulent pipe 46
physical 28
vanishing 6
Elasticity 63, 67, 77, 78, 79, 81, 82, 83, 144
bulk moduli 79
bulk modulus of 67, 77, 78, 79, 81
modulus of 76, 79, 80, 82, 83
Energy 1, 4, 7, 19, 20, 22, 32, 52, 69, 92, 93, 109, 139, 147, 189, 196
and vorticity 1
dissipation rate 92, 93, 139
equation of motion 109
elastic 69
losing 32
Equations 10, 187, 273, 278
integral 187
of turbulent 10
skan 273, 278
Ethyl alcohol glycerin kerosine 81
Euler's equation 274
Expansion corner Wedge flow 172
Experimental turbulent duct 39

F

Favorable pressure gradient 148
Flow 28, 63, 64, 68, 84, 113, 133, 156, 202
conditions 63, 64, 68, 84
rate 113, 133
regulator 202
separation 156
structure 28
Flow velocity 34, 46, 55, 63, 68, 69, 72, 84, 87, 88
profile 46
shear 55
Fluctuating turbulent eddy 57
Fluctuation 2, 3, 17, 23, 24, 26, 27, 36, 49, 51, 55, 57, 100, 101, 136, 177
neglecting pressure 57
random velocity 23
transverse velocity 100
vanish 26, 49, 177
velocities 17, 24, 26

vorticity 3
Fluid 4, 5, 24, 25, 29, 34, 35, 47, 64, 65, 71, 72, 73, 75, 77, 78, 104, 105, 143, 145, 146, 151, 159, 196, 208, 260
compressibility 76, 78, 143, 145
density 77, 146, 260
discharging 196
flows 5, 159
internal energy of 104, 105
kinematic viscosity 260
lump of 35
mechanics 5
motion 47
stagnant 208
viscosity of 34
volume 75, 143
Flush storm runoff 72, 142
Force balance 115
Formula 75, 115, 134
convenient exponential velocity distribution 115, 134
general exponential 75
Free stream 156, 166, 200
Friction 39, 43, 46, 65, 72, 73, 74, 120, 126, 130, 134, 137, 138, 157, 176, 182
dependent 72
factor 43, 120, 130, 134
increased wall 157
neglecting 65
resistance 137
velocity 39, 126, 182
Fundamentals of turbulent flow 3, 5, 7, 9, 11, 13, 15, 17, 19, 21

G

Gaseous mixture 197
Growth 17, 176, 199, 235, 266
deyermine turbulent boundary 176
rapid 17

H

Hammer, steam 64
Heat 2, 157, 197
mass transfer 2
transfer 157, 197
Heuristic treatment 30
Hydropower plant 85
Hypothetical turbulent diffiusion process 10

I

Instantaneous velocity change 64
Integration 61, 181, 182, 183, 200, 246, 263, 267
by partial fraction 61

J

Jet streams 1

K

Karmon's constant 184
Kinematic viscosity 9, 11, 19, 26, 34, 48, 102, 112, 139, 148, 150, 177
eddy 148
molecular 150
Kinetic energy 4, 52, 94, 155
factor 155
instantaneous 52
Kolmogorov's universal 19

L

Laminar 11, 17, 38, 41, 45, 102, 106, 152, 153, 163, 176, 181, 185, 188, 189, 191, 192, 199, 216, 217, 260, 263, 266
and turbulent drag 17
developed 41, 45
dimensional 263
layer, dimensional 260
shear 38
steady 17, 199
Large scale motion 21
Layer 39, 151, 188, 198, 240

control, methods of boundary 188, 198
theory 151
thickness 39, 240
Layer equation 176, 184
turbulent boundary 176
Length hypothesis 37, 269
Length scales 1, 5, 7, 8, 14, 15, 16, 23, 92, 94, 96, 107
convective 1
molecular 5
transverse 15, 16
Liner viscose 41
Liquid 5, 88
turbulence 5
vapor pressure 88
Local velocity profile 166, 190

M

Magnitude 7, 14, 15, 36, 39, 41, 64, 84, 87, 140, 150, 159, 176, 200, 260
analysis 200, 260
layer 176
relative 140, 176
Mass 2, 66, 77, 78, 79, 93, 139, 197
energy dissipation rate/unit 93
densities 78, 79
transfer 197
Mathematical complexity 5
Mechanical vibration 189
Mechanism 33, 35, 100
energy transfer 33
Methods 2, 3, 23, 63, 84, 175, 195, 198
difference computer 175
for controlling transients 63, 84
mathematical 3, 198
of boundary 195
statistical 2, 23
Momentum 1, 2, 17, 18, 23, 24, 25, 28, 29, 33, 34, 35, 102, 124, 158, 159, 160, 223
analysis, integral 223
angular 33
equation perpendicular 102
exchanges 28
fluctuating 28
fluxes 24
transport of 18, 23
Momentum equation 58, 151, 155, 158, 165, 168, 267
integral 151, 158
Momentum thickness 155, 173, 174, 188, 217, 238
single dimensionless 173
Movement 2, 79, 80, 81
longitudinal 79, 80, 81

O

Orifice 86, 87
differential 87
tank 86
Outer turbulent region 185

P

Penstock flow 91
Permissible nose level 99
Piezometric head 72
Pipe 31, 42, 46, 47, 64, 67, 69, 82, 83, 91, 115, 116, 121, 134
equivalent steel 82
rough 46
uniform steel 83
Pipeline 1, 3, 63, 65, 68, 69, 70, 71, 84, 85, 86, 87, 88, 89, 91
profile 84
pressure 89
single 63, 68
reverses 87
Pipe 80, 112, 116
materials 80
radius 116
raduis 112
Plate Reynolds 194
Polyethylene Polystyrene 80
Polyvinyl cloride (PVC) 80, 83
Power law distribution 274
Power plant 85, 90

hydroelectric 84, 85
Prandtl 34, 35, 164
law for laminar 164
theory 34, 35
Prandtl mixing 23, 34, 100, 135, 137
length theory 23, 34, 135
theory 100
length hypothesis 137
Pressure 25, 29, 30, 63, 64, 65, 68, 79, 81, 84, 85, 87, 88, 89, 90, 91, 157, 159, 197, 234
atmospheric 81, 91
difference of 157, 159
hydrodynamic 25
static 234
waterhammer 85
Pressure gradient 34, 41, 139, 140, 148, 156, 173, 176, 179, 191, 197, 204, 205, 226
adverse 140, 179, 197
factor 173, 204, 205, 226
farourable 179
Pressurized transient flow 71
Profile 42, 45, 103, 104, 116, 148, 156, 157, 188, 190, 208, 230
axial velocity 116
boundary layer velocity 188
flat turbulent 45
lacal velocity 148
laminar velocity 190
order polynomial velocity 230
Pump 71, 83, 84, 85, 87, 88, 89, 90, 91
discharging 89
maximum reverse 84

R

Random motion 9
Rapid load, anticipated 90
Rate of supply energy 19
Ratta equation 183
Reinforced 82, 83
concrete pipes 82
plastic pipe 83
Reverse 75, 91, 101
flow 75, 91
order 101
Reynolds 3, 10, 11, 24, 29, 43, 46, 47, 50, 51, 123, 130, 138, 153, 163, 165, 176, 178, 189, 197, 263
and viscous stresses 176
effects 47
equations 24, 50, 51, 178
momentum equation for mean flow 29
number 43, 46, 47, 123, 130, 153, 176, 197, 263
Reynolds stress(es) 23, 24, 29, 30, 32, 34, 47, 50, 140, 142, 176, 178, 179, 181, 182, 267, 268
equation 47
in addition 30
in turbulent flow 34
terms 50, 178

S

Safety valves 88, 89
Scale 5, 18, 19, 20, 21, 22, 107, 139
eddies 21, 157
energy 18, 21
motion 18, 19, 22
smallest 5
time and velocity 20, 139
Separation 3, 62, 138, 155, 156, 157, 188, 195, 196, 197, 199, 266
delay 199, 266
process 157
relegate 196
resist 157
Shear 32, 38, 39, 41, 51, 73
distribution 38
force 73
layer 41
Shear flow 6, 30, 31, 32, 34, 55, 107
in pipes 31
on flat plate 31
Shear stress 34, 37, 115, 118, 134, 136, 149, 173, 176, 226, 242, 256, 257, 258, 259, 263

assumption 263
parameter 226
Sink flow 171
Skan Wedge flows 151, 168
Skin friction 157, 176, 185, 194, 233, 238, 259
coefficient 176, 185, 238
drag 157, 259
force 194
Small scale 6, 18, 21, 96, 97, 100, 106, 107, 139
fluctuation 18
in turbulent 18
rate of energy supply to 106, 107
turbulence 139
Stability analysis 189
Stagnate fluid 208
Stagnation flow 151
State friction losses 72
Statistical analysis 23
Steady state equation 53
Steel 82, 83
bands 83
bars 82
liner 82
Stokes equations flow 176
Stream velocity 17, 39, 165, 256, 260, 278
free 165, 256, 260
local free 278
Stress 4, 36, 39, 104, 112, 113, 116, 125, 126, 133, 144, 174, 181, 263
dimensionless shear 174
dissipative viscous shear 4
eddy 50
pipe flow shear 112
shearing 36
tensile 144
viscous shear 104
wall shear 39, 113, 116, 125, 126, 133, 263
Surge suppressors 89
Surge tank 84, 85, 86, 88, 91
acts 85
equivalent 88
simple 86
Synchronous operation of turbine and pressure 92
System 11, 63, 70, 83, 85, 96
cartesion coordinate 11

T

Tangential forces 157
Taylors theorem 182
Theory 7, 105
asymptotic 7
linearization 7
single 105
Thermal protection 197
Thwaite's method 151
Transient problem 75
Transients 63, 64, 71, 83, 84, 86, 88
hydraulic 63, 71
term hydraulic 64
Transition 188, 189, 191, 192, 193, 194, 217
position 188, 192, 193, 217
process 188, 191
Reynolds number 189, 194
Transport equation 52
Turbine 71, 83, 84, 85, 90, 91, 92, 142
hydraulic 71, 85, 142
flow 91
Turbulence 1, 2, 3, 4, 5, 6, 7, 22, 23, 52, 53, 55, 94, 95, 96, 104, 105, 176
acts 176
diffusivity of 1, 2
models 52
research 3
theory 5
Turbulent 1, 2, 4, 6, 7, 10, 11, 12, 13, 17, 18, 28, 29, 30, 41, 44, 52, 55, 56, 57, 102, 153, 157, 176, 191, 217
agitation 57
atmospheric 13
intensity 6
kinetic energy (TKE) 52, 55, 56
motion 1, 6, 7, 10, 28, 29, 157
nature of 1, 2
pressure 44

stress 29
time scales 6
Transport 30
Turbulent boundary 179, 239
layer flows 179
thickness equation 239
Turbulent flow 42, 182, 188
assumptions 182
solution 42
zone 188
Turbulent fluctuation 27, 28, 36
velocity 27, 36
Turbulent momentum 24, 28, 30
exchange 24
fluxes 24
transfer 28, 30
Turbulent shear 17, 38, 39, 55
flows 17, 55
layer 38
Turbulent velocity 24, 41, 44, 52
fluctuation 24, 52
profile 41, 44

U

Unit width, surface fraction force 192
Unsteady problem 208
Upstream 85, 89, 90, 148, 190, 191
influence 148, 190
reservoir 85

V

Velocity 11, 12, 17, 18, 23, 33, 35, 36, 39, 48, 55, 56, 58, 64, 65, 72, 78, 84, 100, 101, 113, 126, 133, 136, 139, 147, 148, 151, 152, 154, 162, 186, 189, 197, 235, 238, 274
angular 11
centerline 113, 126, 133, 148
change 64
defect law 39
dimensionless 186
distribution 72, 162, 235, 238, 274
disturbance 147, 189
eddies 12
fluctuating 48, 56, 58
gradient effect on boundary layers 33
high supersonic 197
of waterhammer waves 78, 84
shear 55
superimposing 65
transvers 101
uniform 152
Velocity fluctuations 2, 8, 30, 31, 54
in shear flow 31
in shear flow on flat plate 31
Viscose action 152
Viscosity 5, 6, 11, 14, 18, 20, 24, 25, 34, 37, 39, 47, 51, 110, 113, 118, 137, 147, 151, 189, 208
dynamic 25
eddy 5, 24, 34, 37, 113, 118, 137
effect of 32, 151
molecular 6, 11, 14, 39
small 151
Viscous 15, 29, 40, 106, 107, 112, 137, 140, 147, 176, 179, 189
energy loss, rate of 106, 107
force damps 189
force dump 147
linear 137
Vortex stretching 3, 32, 33, 52, 105, 107
productum 107
term 107
Vorticity 1, 3, 7, 21, 22, 33, 52
concentrated 33
fluctuating 3
of small scale 21
stretching 33

W

Wall 37, 38, 39, 46, 55, 56, 58, 59, 67, 95, 78, 81, 96, 102, 103, 104, 111, 112, 137, 138, 143, 144, 149, 157, 197, 241
boundary condition 103
condition 104

law equation 39
plane 241
porous 197
rigid 67
roughness 46, 137
shear 157
smooth 46, 138
thickness 78, 81, 143, 144
Water hammer 63, 64
Waves 63, 64, 66, 69, 71, 78, 79, 82, 83, 147
sound 63
propagate 71
velocity 64, 66, 69, 78, 79, 82, 83, 147
Weisbach 73, 75
formula 73, 75
Williams formulas 75
Wind tunnels 198, 234, 241
laboratory 234

Y

Young's modulus 76, 79, 80

www.ingramcontent.com/pod-product-compliance
Lightning Source LLC
Chambersburg PA
CBHW050812220326
41598CB00006B/190
* 9 7 8 1 6 8 1 0 8 8 1 3 6 *